HOW THE BRAIN WORKS

HOW THE BRAIN WORKS

Editorial Consultant
Rita Carter

Senior Designer
Duncan Turner

Project Art Editors
Amy Child, Mik Gates,
Steve Woosnam-Savage

Illustrators
Mark Clifton,
Phil Gamble, Gus Scott

Managing Art Editor
Michael Duffy

Jacket Designer
Tanya Mehrotra

**Jacket Design
Development Manager**
Sophia MTT

Senior Producer, Pre-production
Andy Hilliard

Art Director
Karen Self

Contributors
Catherine Collin, Tamara Collin, Liam Drew,
Wendy Horobin, Tom Jackson, Katie John, Steve Parker,
Emma Yhnell, Ginny Smith, Nicola Temple, Susan Watt

Lead Senior Editor
Peter Frances

Senior Editor
Rob Houston

Project Editor
Ruth O'Rourke-Jones

Editors
Kate Taylor, Hannah Westlake, Jamie Ambrose,
Camilla Hallinan, Nathan Joyce

Managing Editor
Angeles Gavira Guerrero

Senior Producer
Meskerem Berhane

Publisher
Liz Wheeler

Publishing Director
Jonathan Metcalf

First published in Great Britain in 2020
by Dorling Kindersley Limited
DK, One Embassy Gardens, 8 Viaduct Gardens,
London, SW11 7BW

The authorised representative in the EEA is
Dorling Kindersley Verlag GmbH. Arnulfstr. 124,
80636 Munich, Germany

Copyright © 2020 Dorling Kindersley Limited
A Penguin Random House Company
10 9 8 7
023–315999–Mar/2020

A CIP catalogue record for this book
is available from the British Library.
ISBN: 978-0-2414-0337-2

The information in this book has been compiled as general guidance on the specific subjects addressed.
It is not a substitute and not to be relied on for medical, healthcare or pharmaceutical professional
advice. Please consult your GP before changing, stopping or starting any medical treatment. So far as
the authors are aware the information given is correct and up to date as at October 2019. Practice, laws
and regulations all change and the reader should obtain up to date professional advice on any such
issues. The author and publishers disclaim, as far as the law allows, any liability arising directly or
indirectly from the use or misuse of the information contained in this book.

Printed and bound in China

For the curious
www.dk.com

CONTENTS

THE PHYSICAL BRAIN

BRAIN FUNCTIONS AND THE SENSES

THE
PHYSICAL
BRAIN

What the brain does

The brain is the body's control centre. It coordinates the basic functions required for survival, controls body movements, and processes sensory data. However, it also encodes a lifetime of memories and creates consciousness, imagination, and our sense of self.

DO BRAINS FEEL PAIN?

Despite the fact that it registers pain from around the body, brain tissue has no pain receptors and cannot feel pain itself.

The physical brain

At the largest scale, the human brain appears as a firm, pink-grey solid. It is made mostly from fats (about 60 per cent) and has a density just a little greater than that of water. However, neuroscientists, the people who study the form and function of the brain, see the organ as being constituted from more than 300 separate, although highly interconnected, regions. On a much smaller scale, the brain is made from approximately 160 billion cells, half of which are neurons, or nerve cells, and about half are glia, or support cells of one kind or another (see pp.20–21).

Weight
On average an adult human brain weighs 1.2–1.4 kg (2.6–3.1 lbs), which is approximately 2 per cent of total body weight.

Fat
The brain's dry weight is 60 per cent fat. Much of this fat is present as sheaths coating the connections between neurons.

Water
The brain is 73 per cent water, while the body as a whole is closer to 60 per cent. The average brain contains around one litre (35 fl oz) of water.

Volume
The average volume of a human brain ranges from 1,130 to 1,260 cubic cm (69 to 77 cubic in), although the volume decreases with age.

Grey matter
About 40 per cent of the brain's tissue is grey matter, which is tightly packed nerve-cell bodies.

White matter
Around 60 per cent of the brain's tissue is white matter. This is made from long, wirelike extensions of nerve cells covered in sheaths of fat.

LEFT BRAIN VERSUS RIGHT BRAIN

It is often claimed that one side, or hemisphere, of the brain dominates the other – and that this has an impact on someone's personality. For example, it is sometimes said that logical people use their left brain hemisphere, while artistic (and less logical) people rely on the right side. However, this is an extreme oversimplification. While it is true that the hemispheres are not identical in function – for example, the speech centres are normally on the left – most healthy mental tasks deploy regions on both sides of the brain at the same time.

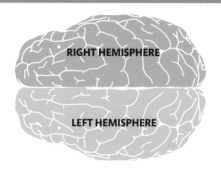

RIGHT HEMISPHERE

LEFT HEMISPHERE

Memory
The brain remembers a bank of semantic knowledge, general facts about the world, as well as a personal record of life history. The function of memory is to aid future survival by encoding useful information from the past.

Movement
To contract, muscles rely on the same kind of electrical impulses that carry nervous signals through the brain and body. All muscle movement is caused by nerve signals, but the conscious brain only has limited control over it.

Emotions
Most theories of emotion suggest that they are preordained modes of behaviour that boost our survival chances when we encounter confusing or dangerous situations. Others suggest emotions are animal instincts leaking through into human consciousness.

What does the brain do?
The relationship between the body and brain has long been a subject of debate for scientists and philosophers. In ancient Egypt, the brain was dismissed as a system for shedding heat, and the heart was the seat of emotion and thought. Although our most significant feelings are still described as heartfelt, neuroscience shows that the brain drives all body activities.

Control
The basic body systems, such as breathing, circulation, digestion, and excretion, are all under the ultimate control of the brain, which seeks to modify their rates to suit the needs of the body.

Communication
A unique feature of the human brain is the speech centres that control the formulation of language and the muscular execution of speech. The brain also uses a predictive system to comprehend what someone else is saying.

Thinking
The brain is where thought and imagination take place. Thinking is a cognitive activity that allows us to interpret the world around us, while our imagination helps us to consider possibilities in the mind without input from the senses.

Sensory experience
Information arriving from all over the body is processed in the brain to create a richly detailed picture of the body's surroundings. The brain filters out a great deal of sensory data deemed irrelevant.

SMOOTHING OUT **ALL THE WRINKLES OF THE BRAIN'S OUTER LAYER** WOULD COVER AN AREA OF ABOUT **2,300 SQUARE CM (2 ½ SQUARE FT)**

The brain in the body

The brain is the primary component of the human body's nervous system, which coordinates the actions of the body with the sensory information it receives.

The nervous system

The two main parts of the nervous system are the central nervous system (CNS) and the peripheral nervous system. The CNS is made up of the brain and the spinal cord, a thick bundle of nerve fibres that runs from the brain in the head to the pelvis. Branching out from this is the peripheral system, a network of nerves that permeates the rest of the body. It is divided according to function: the somatic nervous system handles voluntary movements of the body, while the autonomic nervous system (see opposite) handles involuntary functions.

Spinal nerves

Most peripheral nerves connect to the CNS at the spinal cord and split as they connect. The rear branch carries sensory data to the brain; the forward branch carries motor signals back to the body.

SPINAL COLUMN (REAR VIEW)

Motor nerve

Sensory nerve

SPINAL CORD

SPINAL NERVE

VERTEBRA

Bone vertebra protects spinal cord

CRANIAL NERVES

Within the peripheral system, 12 cranial nerves connect directly to the brain rather than the spinal cord. Most link to the eyes, ears, nose, and tongue and are also involved in facial movements, chewing, and swallowing, but the vagus nerve links directly to the heart, lungs, and digestive organs.

Signals along optic nerve travel directly to brain

Spinal cord

Skull provides protection to brain

Brain

Spinal cord

Permeating the body
The nervous system extends throughout the entire body. It is so complex that all of a body's nerves joined end to end could circle the world two-and-a-half times.

Spinal nerves of peripheral system join spinal cord of central system

Spinal cord runs down back, through vertebrae of spinal column

Peripheral nerves extend through torso and limbs to hands and feet

Sciatic nerve is largest and longest nerve in body

Sensory and motor nerves are often bundled together, separating at their ends

KEY

Central nervous system (CNS)

Peripheral nervous system

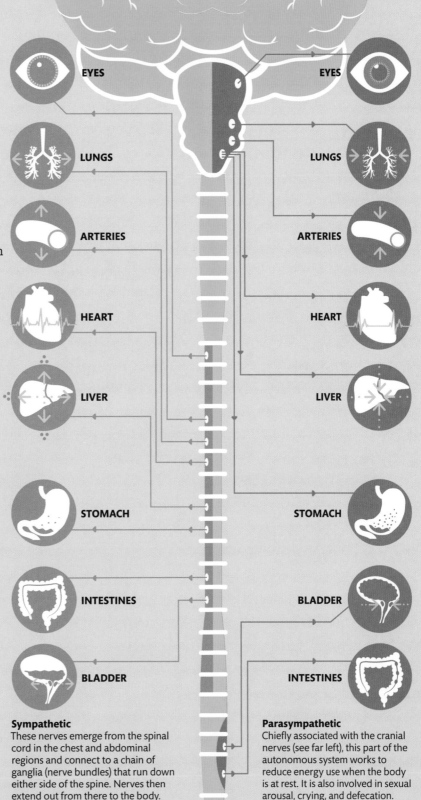

The autonomic nervous system

The involuntary, or autonomic, system maintains the internal conditions of the body by controlling the involuntary muscle in the digestive system and elsewhere, as well as heart and breathing rates, body temperature, and metabolic processes. The autonomic system is divided into two parts. The sympathetic system generally acts to elevate body activity and is involved in the so-called "fight-or-flight" response. The parasympathetic system works in opposition to this, reducing activity to return the body to a "rest-and-digest" state.

EYES

LUNGS

ARTERIES

HEART

LIVER

STOMACH

INTESTINES

BLADDER

EYES

LUNGS

ARTERIES

HEART

LIVER

STOMACH

BLADDER

INTESTINES

THE TOTAL LENGTH OF THE **SOMATIC NERVOUS SYSTEM** IS ABOUT **72 KM (45 MILES)**

Sympathetic
These nerves emerge from the spinal cord in the chest and abdominal regions and connect to a chain of ganglia (nerve bundles) that run down either side of the spine. Nerves then extend out from there to the body.

Parasympathetic
Chiefly associated with the cranial nerves (see far left), this part of the autonomous system works to reduce energy use when the body is at rest. It is also involved in sexual arousal, crying, and defecation.

Human and animal brains

The human brain is one of the defining features of our species. Comparing the human brain with the brains of other animals reveals connections between brain size and intelligence and between an animal's brain anatomy and the way it lives.

Brain sizes

The size of a brain indicates its total processing power. For example, a honeybee's tiny brain contains 1 million neurons, a Nile crocodile's has 80 million, while a human brain has around 80–90 billion neurons. The link with intelligence is clear. However, with larger animals, it is important to compare brain and body size to give a more nuanced indication of cognitive power.

Sizing up
There are two ways to compare brain sizes, by total weight and as a percentage of body weight. The largest brain, at 7.8 kg (17 lb), belongs to the sperm whale, but that is a minute fraction of its 45-tonne (44-ton) body.

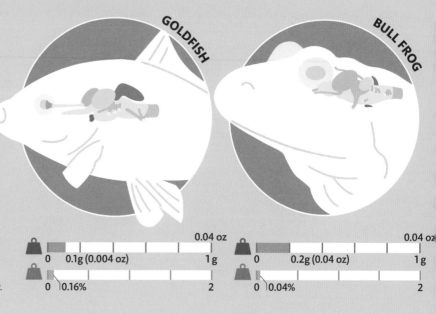

GOLDFISH

BULL FROG

		0.04 oz
🏋 0	0.1g (0.004 oz)	1 g
🏋 0	0.16%	2

		0.04 oz
🏋 0	0.2g (0.04 oz)	1 g
🏋 0	0.04%	2

Brain shapes

All brains are located in the head, in close proximity to the primary sense organs. However, it would be a mistake to visualize animal brains as rudimentary variations, in size and structure, of the human brain. All vertebrate brains follow the same development plan, but anatomies vary widely to match different sensory and behavioural needs. More variety can be seen in the brains of invertebrates, which account for 95 per cent of all animals.

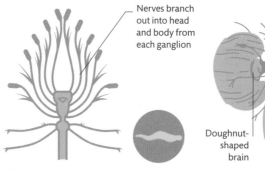

Nerves branch out into head and body from each ganglion

Oesophagus runs through middle of brain

Doughnut-shaped brain

Leech
The 10,000 cells in a leech's nervous system are arranged in chains of cell clusters called ganglia. The brain is a big ganglia, with 350 neurons, located at the front of the body.

Octopus
An octopus's brain contains 500 million neurons. Only a third are located in the head, the rest are in the arms and skin, where they are devoted to sensory and motor controls.

VARYING PROPORTIONS

All mammal brains contain the same components, but they grow in different proportions. A third of the volume of a rat's central nervous system (CNS) is made up of the spinal cord, indicating its reliance on reflex movements. By contrast, the spinal cord is a tenth of a human CNS. Instead, three-quarters is taken up by the cerebrum, which is used for perception and cognition.

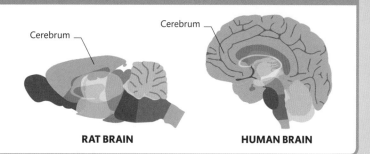

Cerebrum

Cerebrum

RAT BRAIN

HUMAN BRAIN

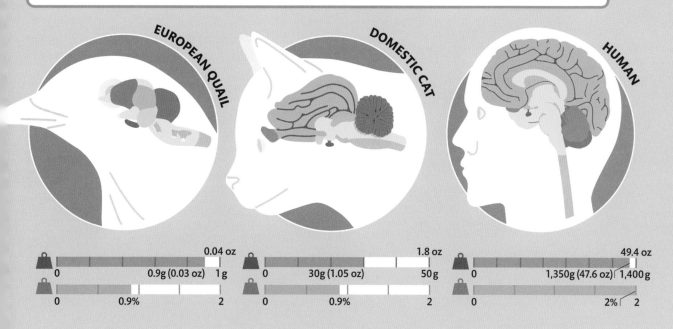

EUROPEAN QUAIL

DOMESTIC CAT

HUMAN

	0.04 oz
0	0.9g (0.03 oz) 1g
0	0.9% 2

	1.8 oz
0	30g (1.05 oz) 50g
0	0.9% 2

	49.4 oz
0	1,350g (47.6 oz) 1,400g
0	2% 2

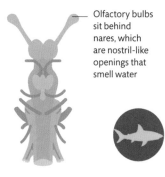

Olfactory bulbs sit behind nares, which are nostril-like openings that smell water

Shark
The brain of a shark is Y-shaped due to the large olfactory bulbs that extend out on either side. The sense of smell is the shark's primary means of tracking prey.

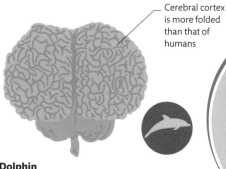

Cerebral cortex is more folded than that of humans

Dolphin
The hearing and vision centres of a dolphin's brain are larger and closer together than in a human brain. It is thought this helps the dolphin create a mental image using its sonar.

DO ALL ANIMALS HAVE A BRAIN?

Sponges have no nerve cells at all, while jellyfish and corals have a netlike nervous system but no central control point.

Protecting the brain

The vital organs are safely secured in the body's core, but because the brain sits in the head at the top of the body it requires its own protection system.

The cranium

The bones of the head are collectively known as the skull, but are more correctly divided into the cranium and the mandible, or jawbone. It is supported by the highest cervical vertebra and the musculature of the neck. The cranium forms a bony case completely surrounding the brain. It is made of 22 bones that steadily fuse together in the early years of life to make a single, rigid structure. Nevertheless, the cranium has around 64 holes, known as foramina, through which nerves and blood vessels pass, and eight air-filled voids, or sinuses, which reduce the weight of the skull.

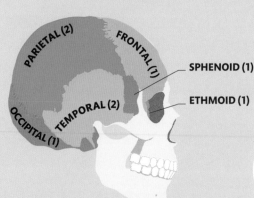

PARIETAL (2)
FRONTAL (1)
SPHENOID (1)
ETHMOID (1)
OCCIPITAL (1)
TEMPORAL (2)

Paired bones
The brain is enclosed by eight large bones, with a pair of parietal and temporal bones forming each side of the cranium. The remaining 14 cranial bones make up the facial skeleton.

Dural sinuses collect oxygen-depleted blood

SUBARACHNOID SPACE

2 **Direction of flow**
CSF flows from the ventricles into the subarachnoid space, where it then moves up and over the front of the brain.

Cerebrospinal fluid

The brain does not come into direct contact the cranium. Instead it is suspended in cerebrospinal fluid (CSF). This clear liquid circulating inside the cranium creates a cushion around the brain to protect it during impacts to the head. In addition, the floating brain does not deform under its own weight, which would otherwise restrict blood flow to the lower internal regions. The exact quantity of CSF also varies to maintain optimal pressure inside the cranium. Reducing the volume of CSF lowers the pressure, which in turn increases the ease with which blood moves through the brain.

WHAT IS WATER ON THE BRAIN?

Also called hydrocephalus, this condition arises when there is too much CSF in the cranium. This puts pressure on the brain and affects its function.

CSF IS CONTINUALLY PRODUCED, AND ALL OF IT **IS REPLACED EVERY 6–8 HOURS**

Meninges and ventricles
The brain is surrounded by three membranes, or meninges: the pia mater, arachnoid mater, and dura mater. The CSF fills cavities called ventricles and circulates around the outside of the brain in the subarachnoid space, which lies between the pia and arachnoid mater.

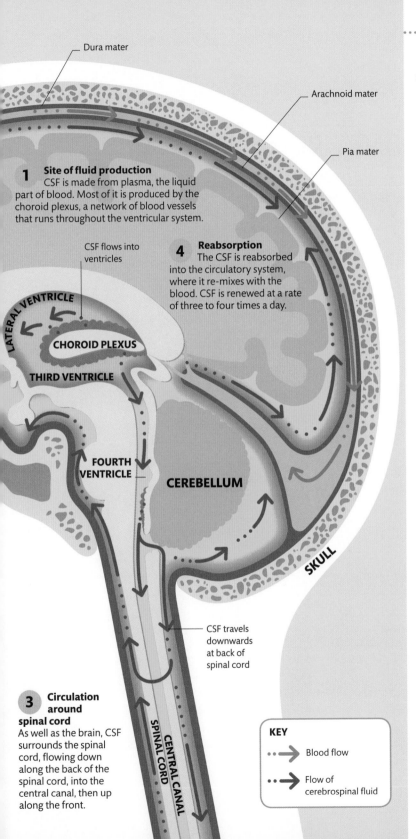

Dura mater

Arachnoid mater

Pia mater

1 Site of fluid production
CSF is made from plasma, the liquid part of blood. Most of it is produced by the choroid plexus, a network of blood vessels that runs throughout the ventricular system.

CSF flows into ventricles

4 Reabsorption
The CSF is reabsorbed into the circulatory system, where it re-mixes with the blood. CSF is renewed at a rate of three to four times a day.

LATERAL VENTRICLE

CHOROID PLEXUS

THIRD VENTRICLE

FOURTH VENTRICLE

CEREBELLUM

SKULL

CSF travels downwards at back of spinal cord

3 Circulation around spinal cord
As well as the brain, CSF surrounds the spinal cord, flowing down along the back of the spinal cord, into the central canal, then up along the front.

CENTRAL CANAL

SPINAL CORD

KEY

```
••• →  Blood flow

••• →  Flow of
        cerebrospinal fluid
```

The blood-brain barrier

Infections from the rest of the body do not ordinarily reach the brain due to a system called the blood–brain barrier. As a general rule, blood capillaries in the rest of the body leak fluid easily (and any viruses and germs it contains) into surrounding tissues through gaps between the cells that form the blood vessel's wall. In the brain, these same cells have a much tighter fit, and the flow of materials between the brain is instead controlled by astrocytes that surround the blood vessels.

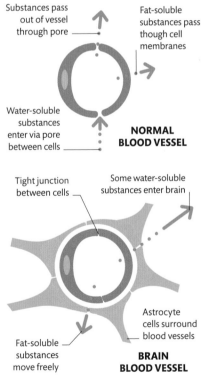

Substances pass out of vessel through pore

Fat-soluble substances pass though cell membranes

Water-soluble substances enter via pore between cells

NORMAL BLOOD VESSEL

Tight junction between cells

Some water-soluble substances enter brain

Fat-soluble substances move freely

Astrocyte cells surround blood vessels

BRAIN BLOOD VESSEL

Selectively permeable
Normal blood vessels allow fluid to pass through easily. However, while oxygen, fat-based hormones, and non-water-soluble materials pass through the blood-brain barrier unhindered, water-soluble items are blocked to prevent them reaching the CSF.

Fuelling the brain

The brain is an energy-hungry organ. Unlike other organs in the body, it is fuelled solely on glucose, a simple sugar that is quick and easy to metabolize.

Blood supply

The heart supplies blood to the whole body, but around one sixth of its total effort is devoted to sending blood up to the brain. Blood reaches the brain by two main arterial routes. The two carotid arteries, one running up each side of the neck, deliver blood to the front of the brain (and the eyes, face, and scalp). The back of the brain is fed by the vertebral arteries, which weave upwards through the spinal column. Deoxygenated blood then accumulates in the cerebral sinuses, which are spaces created by enlarged veins running through the brain. The blood there drains out of the brain and down through the neck via the internal jugular veins.

The vascular system delivers 750 ml (26 fl oz) of blood to the brain every minute, which is equivalent to 50 ml (1.7 fl oz) for every 100 g (3.5 oz) of brain tissue. If that volume drops below about 20 ml (0.7 fl oz), the brain tissue stops working.

DOES FOCUSED CONCENTRATION USE MORE ENERGY?

The brain never stops working, and the overall energy consumption stays more or less the same 24 hours a day.

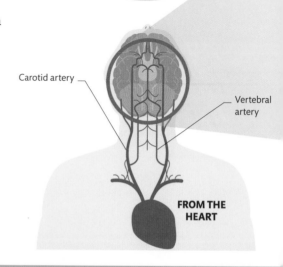

Carotid artery

Vertebral artery

FROM THE HEART

Crossing the blood-brain barrier

The blood-brain barrier is a physical and metabolic barrier between the brain and its blood supply. It offers extra protection against infections, which are hard to combat in the brain using the normal immune system, and could make the brain malfunction in dangerous ways. There are six ways that materials can cross the barrier. Other than that, nothing gets in or out.

Cellular wall

The physical blood-brain barrier is created by the cells that make up the walls of capillaries in the brain. Elsewhere in the body, these are loosely connected, leaving gaps, or loose junctions. In the brain, the cells connect at tight junctions.

Paracellular transport
Water and water-soluble materials, such as salts and ions (charged atoms or molecules), can cross through small gaps between capillary-wall cells.

Diffusion
Cells are surrounded by a fatty membrane, so fat-soluble substances, including oxygen and alcohol, diffuse through the cell.

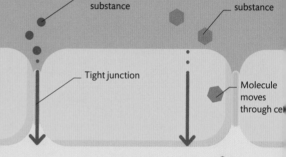

BLOOD VESSEL

BLOOD-BRAIN BARRIER

BRAIN

ASTROCYTE

Water-soluble substance

Fat-soluble substance

Tight junction

Molecule moves through cell

Astrocytes collect material from blood and pass it to neurons

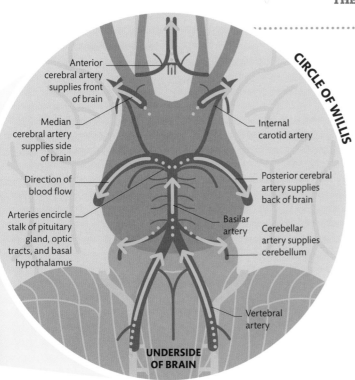

CIRCLE OF WILLIS

Anterior cerebral artery supplies front of brain

Median cerebral artery supplies side of brain

Direction of blood flow

Arteries encircle stalk of pituitary gland, optic tracts, and basal hypothalamus

Internal carotid artery

Posterior cerebral artery supplies back of brain

Basilar artery

Cerebellar artery supplies cerebellum

Vertebral artery

UNDERSIDE OF BRAIN

The Circle of Willis
The carotid and vertebral supplies connect at the base of the brain, via communicating arteries, to create a vascular loop called the Circle of Willis. This feature ensures cerebral blood flow is maintained even if one of the arteries is blocked.

GLUCOSE FUEL

The human brain makes up just two per cent of the body's total weight, but it consumes 20 per cent of its energy. The large human brain is an expensive organ to run, but the benefits of a big, smart brain make it a good investment.

BRAIN SIZE: 2%

BRAIN'S ENERGY NEEDS: 20%

THE BODY'S **ENTIRE SUPPLY OF BLOOD IS** PUMPED THROUGH THE BRAIN **EVERY 7 MINUTES**

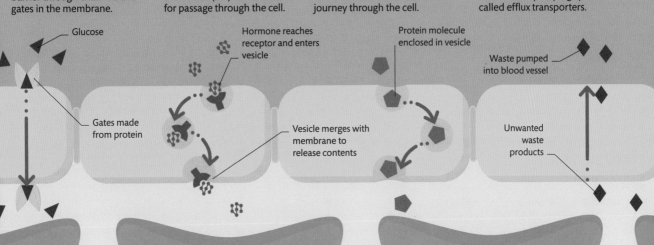

Protein transporters
Glucose and other essential molecules are actively moved across the barrier through channels and gates in the membrane.

Receptors
Hormones and similar substances are picked up by receptors. They are enclosed in a vesicle (sac) of membrane for passage through the cell.

Transcytosis
Large proteins, which are too big to pass through channels, are absorbed by the membrane and enclosed in a vesicle for its journey through the cell.

Active efflux
When unwanted materials diffuse through the blood-brain barrier, they are removed by a biochemical pumping system called efflux transporters.

Glucose

Gates made from protein

Hormone reaches receptor and enters vesicle

Vesicle merges with membrane to release contents

Protein molecule enclosed in vesicle

Waste pumped into blood vessel

Unwanted waste products

Brain cells

The brain and the rest of the nervous system contains a network of cells called neurons. The role of neurons is to carry nerve signals through the brain and body as electrical pulses.

Neurons

Most neurons have a distinctive branched shape with dozens of filaments, only a few millionths of a metre thick, extending from the cell body towards nearby cells. Branches called dendrites bring signals into the cell, while a single branch, called the axon, passes the signal to the next neuron. In most cases there is no physical connection between neurons. Instead, there is a tiny gap, called the synapse, where electrical signals stop. Communication between cells is carried out by the exchange of chemicals, called neurotransmitters (see pp.22–23). However, some neurons are effectively physically connected and do not need a neurotransmitter to exchange signals.

GREY MATTER

The brain is divided into grey and white matter. Grey matter is made of neuron cell bodies, common in the surface of the brain. White matter is made of these neurons' myelinated axons bundled into tracts. They run through the middle of the brain and down the spinal cord.

WHITE MATTER
GREY MATTER

Types of neuron

There are several types of neuron, with different combinations of axons and dendrites. Two common types, bipolar and multipolar neurons, are each suited to particular tasks. Another type of neuron, the unipolar neuron, only appears in embyros.

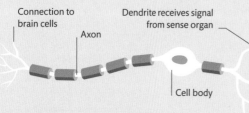

Connection to brain cells

Axon

Dendrite receives signal from sense organ

Cell body

Bipolar neuron

This type of neuron has one dendrite and one axon. It transmits specialized information from the body's major sense organs.

Dendrites act like antennae to collect signals from neighbouring nerve cells

Electrical pulse jumps from one myelin segment to the next, speeding up nerve signal

Axons can be several centimetres long

Dendrites are shorter than axons, usually only up to 50 millionths of a metre long

AXON

Axon delivers signal from neighbouring cell

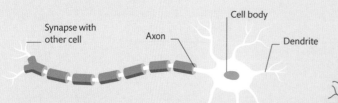

Synapse with other cell

Axon

Cell body

Dendrite

Multipolar neuron

Most brain cells are multipolar. They have multiple dendrites connecting to hundreds, even thousands, of other cells.

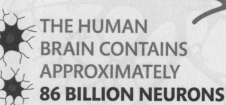

THE HUMAN BRAIN CONTAINS APPROXIMATELY **86 BILLION NEURONS**

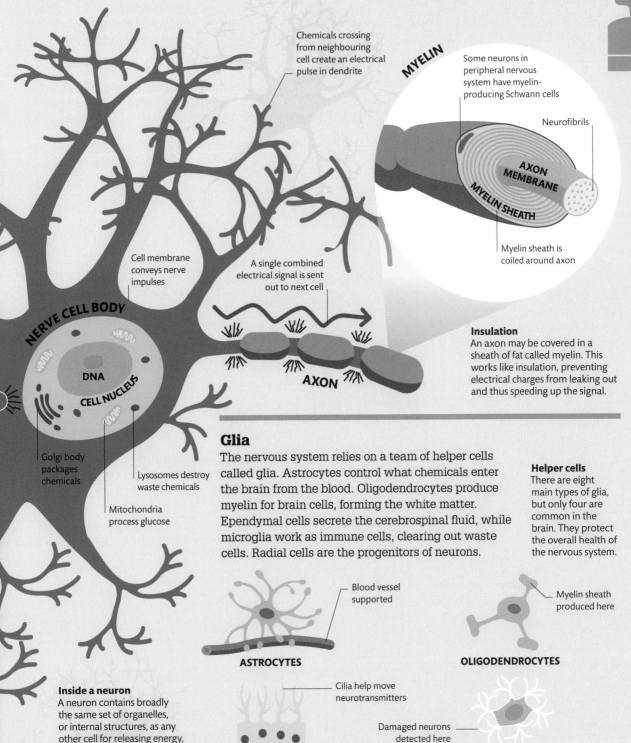

Chemicals crossing from neighbouring cell create an electrical pulse in dendrite

MYELIN

Some neurons in peripheral nervous system have myelin-producing Schwann cells

Neurofibrils

AXON MEMBRANE

MYELIN SHEATH

Myelin sheath is coiled around axon

Cell membrane conveys nerve impulses

A single combined electrical signal is sent out to next cell

NERVE CELL BODY

DNA

CELL NUCLEUS

AXON

Insulation
An axon may be covered in a sheath of fat called myelin. This works like insulation, preventing electrical charges from leaking out and thus speeding up the signal.

Golgi body packages chemicals

Lysosomes destroy waste chemicals

Mitochondria process glucose

Glia

The nervous system relies on a team of helper cells called glia. Astrocytes control what chemicals enter the brain from the blood. Oligodendrocytes produce myelin for brain cells, forming the white matter. Ependymal cells secrete the cerebrospinal fluid, while microglia work as immune cells, clearing out waste cells. Radial cells are the progenitors of neurons.

Helper cells
There are eight main types of glia, but only four are common in the brain. They protect the overall health of the nervous system.

Blood vessel supported

Myelin sheath produced here

ASTROCYTES

OLIGODENDROCYTES

Inside a neuron
A neuron contains broadly the same set of organelles, or internal structures, as any other cell for releasing energy, making proteins, and managing genetic material.

Cilia help move neurotransmitters

Damaged neurons detected here

EPENDYMAL CELLS

MICROGLIA

Nerve signals

The brain and nervous system work by sending signals through cells as pulses of electrical charge and between cells either by using chemical messengers called neurotransmitters or by electric charge.

Action potential

Neurons signal by creating an action potential – a surge of electricity created by sodium and potassium ions crossing the cell's membrane. It travels down the axon and stimulates receptors on dendrites of neighbouring cells. The junction between cells is called a synapse. In many neurons, the charge is carried over a minute gap between axon and dendrite by chemicals, called neurotransmitters, released from the tip of the axon. These junctions are known as chemical synapses. The signal may cause the neighbouring neuron to fire, or it may stop it from firing.

HOW DOES A NERVE COMMUNICATE DIFFERENT INFORMATION?

Receiving cells have different types of receptors, which respond to different neurotransmitters. The "message" differs according to which neurotransmitters are sent and received and in what quantities.

SOME **NERVE IMPULSES** TRAVEL FASTER THAN **100 M (330 FT) PER SECOND**

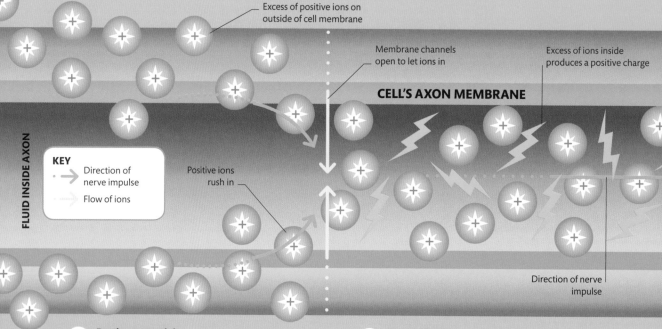

Excess of positive ions on outside of cell membrane

Membrane channels open to let ions in

Excess of ions inside produces a positive charge

CELL'S AXON MEMBRANE

FLUID INSIDE AXON

KEY
- → Direction of nerve impulse
- → Flow of ions

Positive ions rush in

Direction of nerve impulse

1 Resting potential
When the neuron is at rest, there are more positive ions outside the membrane than inside. This causes a difference in polarization, or electrical potential, across the membrane called the resting potential. The difference is about -70 millivolts, meaning the outside is positive.

2 Depolarization
Chemical changes from the cell body allow positive ions to flood into the cell through the membrane. That reverses the polarization of the axon, making the potential difference +30 millivolts.

NERVE AGENTS

Chemical weapons, like novichok and sarin, work by interfering with how neurotransmitters behave at the synapse. Nerve agents can be inhaled or act on contact with skin. They prevent the synapse from clearing away used acetylcholine, which is involved in the control of muscles. As a result, muscles, including those used by the heart and lungs, are paralysed.

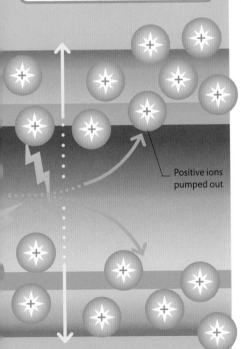

3 Repolarization
The depolarization of a section of the axon causes the neighbouring section to undergo the same process. Meanwhile, the cell pumps out positive ions to repolarize the membrane back to the resting potential.

Positive ions pumped out

Synapses

Some neurons do not share a physical connection. Instead they meet at a cellular structure, called a synapse, where there is a gap of 40 billionths of a metre, known as the synaptic cleft, between the axon of one neuron (the presynaptic cell) and the dendrite of another (the postsynaptic cell). Any coded signal carried by electrical pulses is converted into a chemical message at the tip, or terminal, of the axon. The messages take the form of one of several molecules called neurotransmitters (see p.24), which pass across the synaptic cleft to be received by the dendrite. Other neurons have electrical synapses rather than chemical synapses. These are effectively physically connected and do not need a neurotransmitter to carry electrical charge between them.

1 Chemical store
Neurotransmitters are manufactured in the cell body of the neuron. They travel along the axon to the terminal, where they are parcelled up into membranous sacs, or vesicles. At this stage, the terminal's membrane carries the same electrical potential as the rest of the axon.

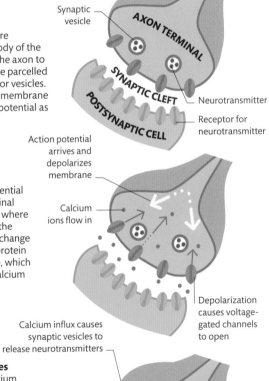

Synaptic vesicle

AXON TERMINAL

SYNAPTIC CLEFT

POSTSYNAPTIC CELL

Neurotransmitter

Receptor for neurotransmitter

Action potential arrives and depolarizes membrane

Calcium ions flow in

Depolarization causes voltage-gated channels to open

2 Signal received
When an action potential surges down the axon, its final destination is the terminal, where it temporarily depolarizes the membrane. This electrical change has the effect of opening protein channels in the membrane, which allow positively charged calcium ions to flood into the cell.

Calcium influx causes synaptic vesicles to release neurotransmitters

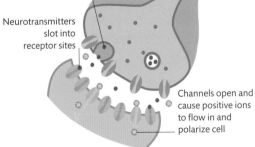

Neurotransmitters slot into receptor sites

Channels open and cause positive ions to flow in and polarize cell

3 Releasing messages
The presence of calcium within the cell sets off a complex process that moves the vesicles to the cell membrane. Once there, the vesicles release neurotransmitters into the cleft. Some diffuse across the gap to be picked up by receptors on the dendrite. The neurotransmitters may stimulate an action potential to form in that dendrite, or they may inhibit one from forming.

Brain chemicals

While communication in the brain relies on electric pulses flashing along wirelike nerve cells, the activity of these cells – and the mental and physical states they induce – are heavily influenced by chemicals called neurotransmitters.

Neurotransmitters

Neurotransmitters are active at the synapse, the tiny gap between the axon of one cell and a dendrite of another (see p.23). Some neurotransmitters are excitatory, meaning that they help continue the transmission of an electrical nerve impulse to the receiving dendrite. Inhibitory neurotransmitters have the opposite effect. They create an elevated negative electrical charge, which stops the transmission of the nerve impulse by preventing depolarization from taking place. Other neurotransmitters, called neuromodulators, modulate the activity of other neurons in the brain. Neuromodulators spend more time at the synapse, so they have more time to affect neurons.

Drugs

Chemicals that change mental and physical states, both legal and illegal, generally act by interacting with a neurotransmitter. For example, caffeine blocks adenosine receptors, which has the effect of increasing wakefulness. Alcohol stimulates GABA receptors and inhibits glutamate, both inhibiting neural activity in general. Nicotine activates the receptors for acetylcholine, which has several effects, including an increase in attention as well as elevated heart rate and blood pressure. Both alcohol and nicotine have been linked to an elevation of dopamine in the brain, which is what leads to their highly addictive qualities.

TYPES OF NEUROTRANSMITTER

There are at least 100 neurotransmitters, some of which are listed below. Whether a neurotransmitter is excitatory or inhibitory is determined by the presynaptic neuron that released it.

NEUROTRANSMITTER CHEMICAL NAME	USUAL POSTSYNAPTIC EFFECT
Acetylcholine	Mostly excitatory
Gamma-aminobutyric acid (GABA)	Inhibitory
Glutamate	Excitatory
Dopamine	Excitatory and inhibitory
Noradrenaline	Mostly excitatory
Serotonin	Inhibitory
Histamine	Excitatory

TYPE OF DRUG		EFFECTS
	Agonist	A brain chemical that stimulates the receptor associated with a particular neurotransmitter, elevating its effects.
	Antagonist	A molecule that does the opposite of an agonist, by inhibiting the action of receptors associated with a neurotransmitter.
	Reuptake inhibitor	A chemical that stops a neurotransmitter being reabsorbed by the sending neuron, thus causing an agonistic response.

BLACK WIDOW SPIDER VENOM INCREASES LEVELS OF THE NEUROTRANSMITTER ACETYLCHOLINE, WHICH CAUSES MUSCLE SPASMS

THE LONG-TERM EFFECTS OF ALCOHOL

Drinking large volumes of alcohol over a long period alters mood, arousal, behaviour, and neuropsychological functioning. Alcohol's depressant effect both excites GABA and inhibits glutamate, decreasing brain activity. It also triggers the brain's reward centres by releasing dopamine, in some cases leading to addiction.

KEY
- Dopamine
- Cocaine

Dopamine and cocaine
The effects of cocaine are a product of its effects on the neurotransmitter dopamine at synapses in the brain.

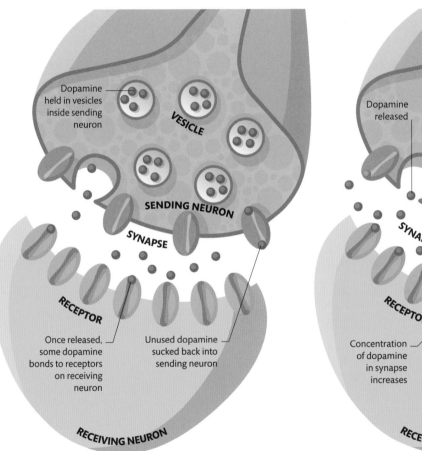

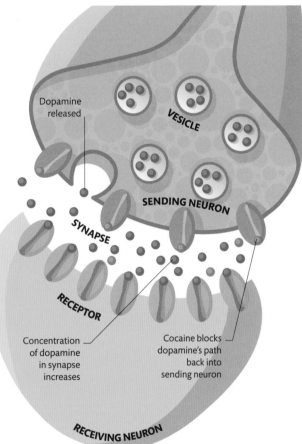

Dopamine held in vesicles inside sending neuron

VESICLE

SENDING NEURON

SYNAPSE

RECEPTOR

Once released, some dopamine bonds to receptors on receiving neuron

Unused dopamine sucked back into sending neuron

RECEIVING NEURON

Dopamine released

VESICLE

SENDING NEURON

SYNAPSE

RECEPTOR

Concentration of dopamine in synapse increases

Cocaine blocks dopamine's path back into sending neuron

RECEIVING NEURON

Normal dopamine levels
Dopamine is a neurotransmitter associated with feeling pleasure. It creates a drive to repeat certain behaviours that trigger feelings of reward, perhaps leading to addiction. While some dopamine molecules bind to receptors on the receiving neuron, unused dopamine is recycled by being pumped back into the sending neuron and parcelled up again.

With use of cocaine
Cocaine molecules are reuptake inhibitors of dopamine. When dopamine is released, it moves into the synapse and binds to receptors on the receiving neuron as normal. However, the cocaine has blocked the reuptake pumps that recycle the dopamine, so the neurotransmitter accumulates in a higher concentration, increasing its effects on the receiving neuron.

Networks in the brain

The patterns of nerve-cell connections in the human brain are believed to influence how it processes sensory perceptions, performs cognitive tasks, and stores memories.

Wiring the brain

The dominant theory of how the brain remembers and learns can be summed up by the phrase "the cells that fire together, wire together". It suggests that repeated communication between cells creates stronger connections between them, and a network of cells emerges in the brain that is associated with a specific mental process – such as a movement, a thought, or even a memory (see pp.136–37).

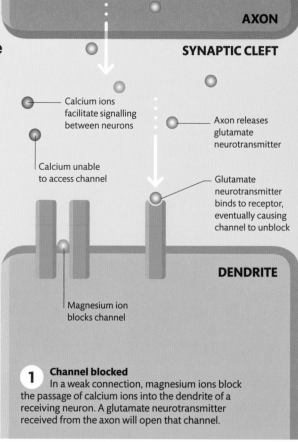

AXON

SYNAPTIC CLEFT

Calcium ions facilitate signalling between neurons

Axon releases glutamate neurotransmitter

Calcium unable to access channel

Glutamate neurotransmitter binds to receptor, eventually causing channel to unblock

DENDRITE

Magnesium ion blocks channel

KEY

- Magnesium ion
- Calcium ion
- Glutamate neurotransmitter
- Channel
- Glutamate receptor

Synaptic weight

Little-used connections have channels blocked by magnesium ions. As the strength of a connection between two neurons in a network increases, the channel is unblocked, and the number of receptors at the synapse increases.

1 Channel blocked

In a weak connection, magnesium ions block the passage of calcium ions into the dendrite of a receiving neuron. A glutamate neurotransmitter received from the axon will open that channel.

Neuroplasticity

The networks of the brain are not fixed but seem to change and adapt in accordance with mental and physical processes. This means that old circuits associated with one memory or a skill that is no longer in use fade in strength as the brain devotes attention to another and forms a new network with other cells. Neuroscientists say the brain is plastic, meaning its cells and the connections between them can be reformed many times over as required. Neuroplasticity allows brains to recover abilities lost due to brain damage.

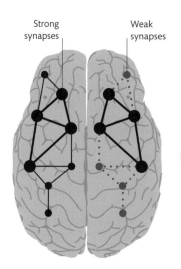

Strong synapses

Weak synapses

BRAIN PATHWAYS

WHAT IS THE BRAIN'S DEFAULT MODE NETWORK?

It is a group of brain regions that show low activity levels when engaged in a task such as paying attention, but high activity levels when awake and not engaged in a specific mental task.

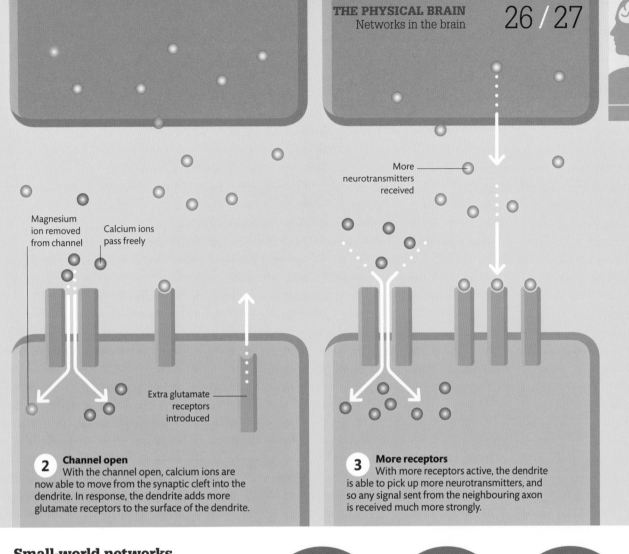

More
neurotransmitters
received

Magnesium
ion removed
from channel

Calcium ions
pass freely

Extra glutamate
receptors
introduced

2 Channel open
With the channel open, calcium ions are now able to move from the synaptic cleft into the dendrite. In response, the dendrite adds more glutamate receptors to the surface of the dendrite.

3 More receptors
With more receptors active, the dendrite is able to pick up more neurotransmitters, and so any signal sent from the neighbouring axon is received much more strongly.

Small-world networks

Brain cells are not connected in a regular pattern, nor are they in a random network. Instead, many of them exhibit a form of small-world network, where cells are seldom connected to their immediate neighbours but to nearby ones. This way of networking allows each cell to, on average, connect to any other in the smallest number of steps.

Random
A random network is good at making long-distance connections but poor at linking nearby cells.

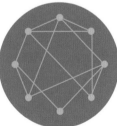

Small-world
Small-world networks have good local and distance connections. Every cell is more closely linked than in the other two systems.

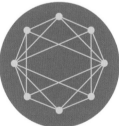

Lattice
By connecting every cell to its neighbours, this network has reduced scope to make long-distance connections.

IT IS ESTIMATED THAT THE **HUMAN BRAIN CONTAINS 100 TRILLION CONNECTIONS** BETWEEN ITS **86 BILLION NEURONS**

Brain anatomy

The brain is a complex mass of soft tissue composed almost entirely of neurons, glial cells (see p.21), and blood vessels, which are grouped into an outer layer, the cortex, and other specialized structures.

Divisions of the brain

The brain is divided into three unequal parts: the forebrain, midbrain, and hindbrain. These divisions are based on how they develop in the embryonic brain, but they also reflect differences in function. In the human brain, the forebrain dominates, making up nearly 90 per cent of the brain by weight. It is associated with sensory perception and higher executive functions. The midbrain and hindbrain below it are more involved with the basic bodily functions that determine survival, such as sleep and alertness.

SPINAL NERVES

There are 31 pairs of spinal nerves that branch out from the spinal cord above each vertebral bone, named after the parts of the spine to which they connect. They relay signals between the brain and sensory organs, muscles, and glands.

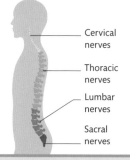

Cervical nerves

Thoracic nerves

Lumbar nerves

Sacral nerves

Surface layer of forebrain, known as grey matter, is made from unsheathed neurons

Tracts of white matter – neurons sheathed with fatty myelin

CORTEX

CEREBRUM

GREY MATTER

CORPUS CALLOSUM

HIPPOCAMPUS

AMYGDALA

THALAMUS

MIDBRAIN

CEREBELLUM

PONS

MEDULLA

BRAINSTEM

SPINAL CORD

Midbrain

The smallest brain section, this is associated with the sleep-wake cycle, thermoregulation (control of body temperature), and visual reflexes, such as the rapid eye movements that scan complex scenes automatically. The substantia nigra, which is a region associated with planning smooth muscle control, is in the midbrain.

Hindbrain

Made up of the cerebellum at the lower rear of the brain and the brainstem, which connects to the spinal cord, the hindbrain is the most primitive part of the brain. The genes that control its development evolved around 560 million years ago.

Direct connections to all three sections of brain are carried in spinal cord

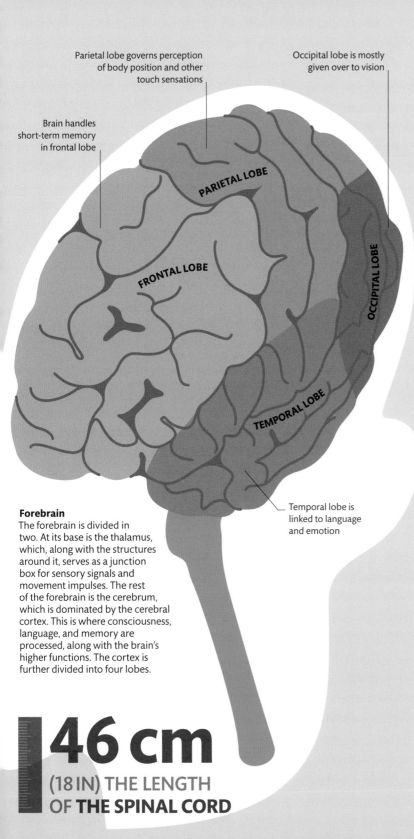

Parietal lobe governs perception of body position and other touch sensations

Occipital lobe is mostly given over to vision

Brain handles short-term memory in frontal lobe

PARIETAL LOBE

FRONTAL LOBE

OCCIPITAL LOBE

TEMPORAL LOBE

Temporal lobe is linked to language and emotion

Forebrain

The forebrain is divided in two. At its base is the thalamus, which, along with the structures around it, serves as a junction box for sensory signals and movement impulses. The rest of the forebrain is the cerebrum, which is dominated by the cerebral cortex. This is where consciousness, language, and memory are processed, along with the brain's higher functions. The cortex is further divided into four lobes.

46 cm
(18 IN) THE LENGTH
OF **THE SPINAL CORD**

Hemispheres

The cerebrum forms in two halves, or hemispheres, which are divided laterally by a gap called the longitudinal fissure. Nevertheless the hemispheres share an extensive connection via the corpus callosum. Each hemisphere is a mirror image of the other, although not all functions are performed by both sides (see p.10). For example, speech centres tend to be on the left side.

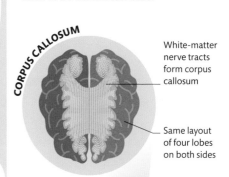

CORPUS CALLOSUM

White-matter nerve tracts form corpus callosum

Same layout of four lobes on both sides

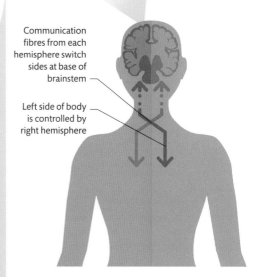

Communication fibres from each hemisphere switch sides at base of brainstem

Left side of body is controlled by right hemisphere

Left and right

The brain and the body are connected contralaterally, meaning that the left brain hemisphere handles the sensations and movements of the right side of the body and vice versa.

The cortex

The cortex is the thin outer layer that forms the brain's visible surface. It has several important functions, including handling sensory data and language processing. It also works to generate our conscious experience of the world.

A functional map

The cortex is a multi-layered coating of neurons, with their cell bodies at the top. Neuroscientists divide it into areas where the cells appear to work together to perform a particular function. There are different ways to reveal this information: through the location of brain damage linked to the loss of a brain function; tracking the connections between cells; and through scans of live brain activity.

WHAT IS PHRENOLOGY?

A 19th-century pseudoscience, in which the shape of the head was linked to brain structure, specific abilities, and personality.

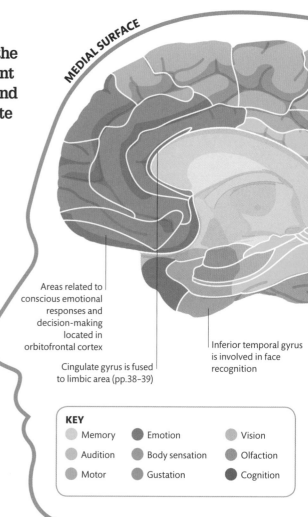

MEDIAL SURFACE

Areas related to conscious emotional responses and decision-making located in orbitofrontal cortex

Cingulate gyrus is fused to limbic area (pp.38–39)

Inferior temporal gyrus is involved in face recognition

KEY

● Memory	● Emotion	● Vision
● Audition	● Body sensation	● Olfaction
● Motor	● Gustation	● Cognition

Folds and grooves

The cerebral cortex is a feature of all mammal brains, but the human brain is distinctive because of its highly folded appearance. The many folds increase the total surface area of the cortex, thereby providing more room for larger cortical areas. The groove in a fold is called a sulcus, and the ridge is called a gyrus. Every human brain has the same pattern of gyri and sulci, which neuroscientists employ to describe specific locations in the cortex.

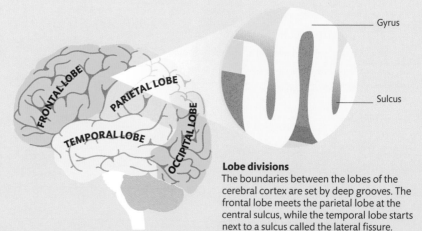

FRONTAL LOBE

PARIETAL LOBE

TEMPORAL LOBE

OCCIPITAL LOBE

Gyrus

Sulcus

Lobe divisions
The boundaries between the lobes of the cerebral cortex are set by deep grooves. The frontal lobe meets the parietal lobe at the central sulcus, while the temporal lobe starts next to a sulcus called the lateral fissure.

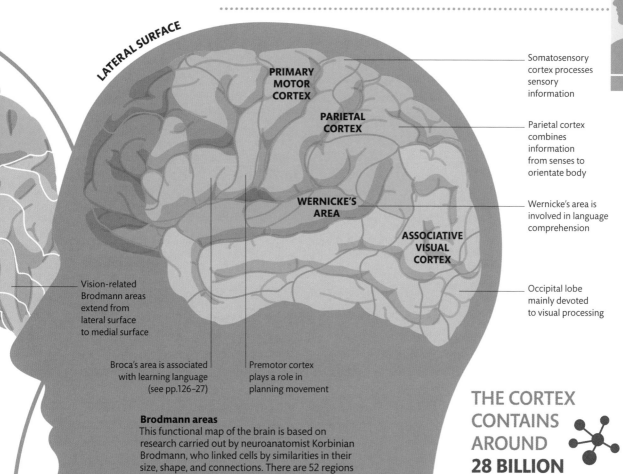

LATERAL SURFACE

PRIMARY MOTOR CORTEX

PARIETAL CORTEX

WERNICKE'S AREA

ASSOCIATIVE VISUAL CORTEX

Somatosensory cortex processes sensory information

Parietal cortex combines information from senses to orientate body

Wernicke's area is involved in language comprehension

Occipital lobe mainly devoted to visual processing

Vision-related Brodmann areas extend from lateral surface to medial surface

Broca's area is associated with learning language (see pp.126–27)

Premotor cortex plays a role in planning movement

Brodmann areas
This functional map of the brain is based on research carried out by neuroanatomist Korbinian Brodmann, who linked cells by similarities in their size, shape, and connections. There are 52 regions in total, and each one can be associated with one or more approximate functions.

THE CORTEX CONTAINS AROUND 28 BILLION NEURONS

Cell structure

The cells of the human cortex are arranged in six layers, with a total thickness of 2.5 mm (0.09 in). Each layer contains different types of cortical neurons that receive and send signals to other areas of the cortex and the rest of the brain. The constant relaying of data keeps all parts of the brain aware of what is going on elsewhere. Some of the more primitive parts of the human brain, such as the hippocampal fold, only have three layers.

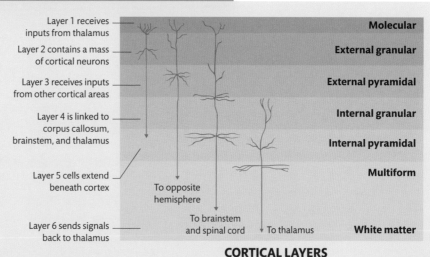

Layer 1 receives inputs from thalamus

Layer 2 contains a mass of cortical neurons

Layer 3 receives inputs from other cortical areas

Layer 4 is linked to corpus callosum, brainstem, and thalamus

Layer 5 cells extend beneath cortex

Layer 6 sends signals back to thalamus

To opposite hemisphere

To brainstem and spinal cord

To thalamus

Molecular

External granular

External pyramidal

Internal granular

Internal pyramidal

Multiform

White matter

CORTICAL LAYERS

Nuclei of the brain

In brain anatomy, a nucleus is a cluster of nerve cells that have a discernible set of functions and are connected to each other by tracts of white matter.

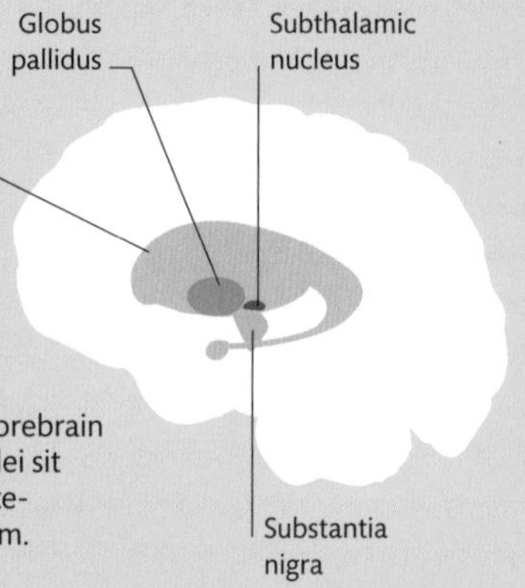

The basal ganglia and other nuclei

An important group of nuclei collectively known as the basal ganglia sit within the forebrain and have strong links with the thalamus and brainstem. They are associated with learning, motor control, and emotional responses. All cranial nerves connect to the brain at a nucleus (often two, one for sensory inputs and another for motor outputs). Other brain nuclei include the hypothalamus (see p.34), hippocampus (see pp.38–39), pons, and medulla (see p.36).

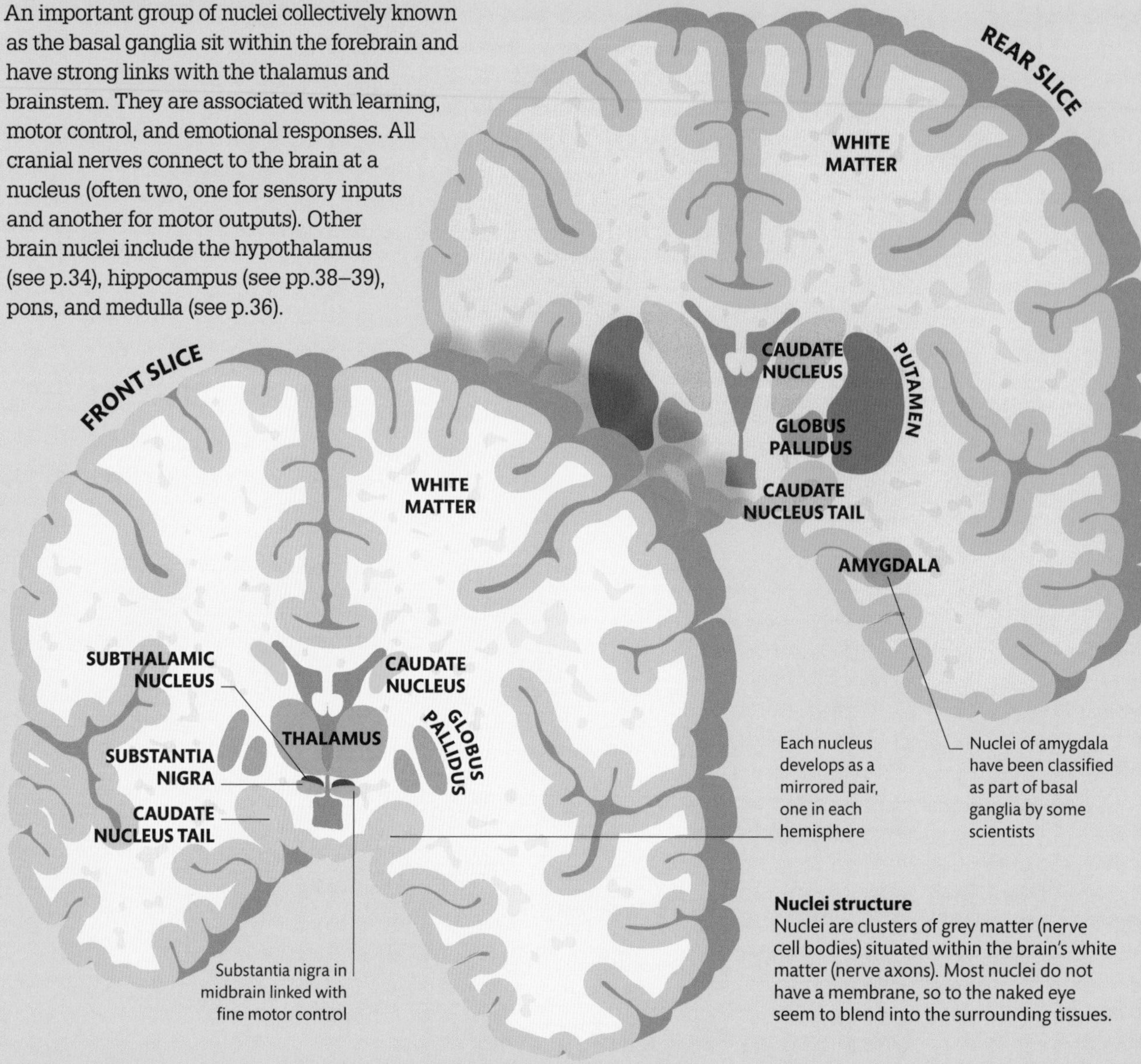

Central location

Most of the basal ganglia are positioned at the base of the forebrain around the thalamus. The nuclei sit within a region filled with white-matter tracts called the striatum.

Globus pallidus

Subthalamic nucleus

Caudate nucleus

Substantia nigra

REAR SLICE

WHITE MATTER

CAUDATE NUCLEUS

PUTAMEN

GLOBUS PALLIDUS

CAUDATE NUCLEUS TAIL

AMYGDALA

FRONT SLICE

WHITE MATTER

SUBTHALAMIC NUCLEUS

CAUDATE NUCLEUS

THALAMUS

GLOBUS PALLIDUS

SUBSTANTIA NIGRA

CAUDATE NUCLEUS TAIL

Each nucleus develops as a mirrored pair, one in each hemisphere

Nuclei of amygdala have been classified as part of basal ganglia by some scientists

Substantia nigra in midbrain linked with fine motor control

Nuclei structure

Nuclei are clusters of grey matter (nerve cell bodies) situated within the brain's white matter (nerve axons). Most nuclei do not have a membrane, so to the naked eye seem to blend into the surrounding tissues.

WHAT NUCLEI ARE LOCATED IN THE BRAINSTEM?

The brainstem contains 10 of the 12 pairs of cranial nuclei. They provide motor and sensory function to the tongue, larynx, facial muscles, and more.

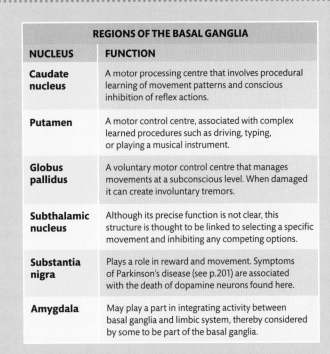

REGIONS OF THE BASAL GANGLIA	
NUCLEUS	**FUNCTION**
Caudate nucleus	A motor processing centre that involves procedural learning of movement patterns and conscious inhibition of reflex actions.
Putamen	A motor control centre, associated with complex learned procedures such as driving, typing, or playing a musical instrument.
Globus pallidus	A voluntary motor control centre that manages movements at a subconscious level. When damaged it can create involuntary tremors.
Subthalamic nucleus	Although its precise function is not clear, this structure is thought to be linked to selecting a specific movement and inhibiting any competing options.
Substantia nigra	Plays a role in reward and movement. Symptoms of Parkinson's disease (see p.201) are associated with the death of dopamine neurons found here.
Amygdala	May play a part in integrating activity between basal ganglia and limbic system, thereby considered by some to be part of the basal ganglia.

THE BRAIN HAS MORE THAN **30 SETS** OF NUCLEI, MOSTLY PAIRED **LEFT** AND **RIGHT**

Action selection

The basal ganglia have an important role in filtering out the noise of competing commands coming from the cortex and elsewhere in the forebrain. This process is called action selection, and it occurs entirely subconsciously through a series of pathways through the basal ganglia. Generally, these pathways block or inhibit a specific action by having the thalamus loop the signal back to the start point. However, when the pathway is silent, the action goes ahead.

Basal ganglia loops

The route of the pathway depends on the source of the inputs from the cortex or elsewhere in the forebrain. There are three main pathways and each one is able to inhibit or select an action. The motor loop connects to the main movement control centre, the prefrontal loop carries input from executive regions of the brain, while the limbic loop is governed by emotional stimuli.

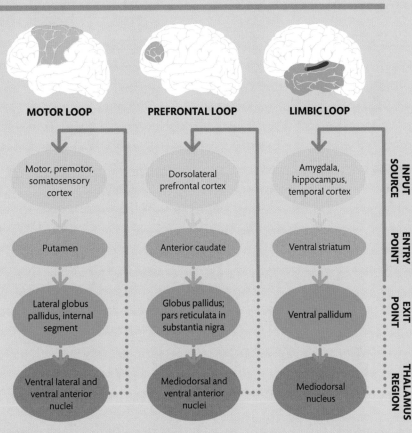

	MOTOR LOOP	PREFRONTAL LOOP	LIMBIC LOOP
INPUT SOURCE	Motor, premotor, somatosensory cortex	Dorsolateral prefrontal cortex	Amygdala, hippocampus, temporal cortex
ENTRY POINT	Putamen	Anterior caudate	Ventral striatum
EXIT POINT	Lateral globus pallidus, internal segment	Globus pallidus; pars reticulata in substantia nigra	Ventral pallidum
THALAMUS REGION	Ventral lateral and ventral anterior nuclei	Mediodorsal and ventral anterior nuclei	Mediodorsal nucleus

Hypothalamus, thalamus, and pituitary gland

The thalamus and the structures around it sit at the centre of the brain. They act as relay stations between the forebrain and the brainstem, also forming a link to the rest of the body.

The hypothalamus

This small region under the forward region of the thalamus is the main interface between the brain and the hormone, or endocrine, system. It does this by releasing hormones directly into the bloodstream, or by sending commands to the pituitary gland to release them. The hypothalamus has a role in growth, homeostasis (maintaining optimal body conditions), and significant behaviours such as eating and sex. This makes it responsive to many different stimuli.

WHAT GLANDS DOES THE PITUITARY GLAND CONTROL?

The pituitary gland is a master gland that controls the thyroid gland, adrenal gland, ovaries, and testes. However, it receives its instructions from the hypothalamus.

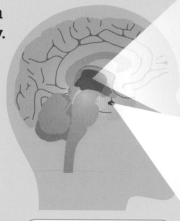

KEY

- Thalamus
- Hypothalamus
- Pituitary gland

THE EPITHALAMUS

This small region covers the top of the thalamus. It contains various nerve tracts that form a connection between the forebrain and midbrain. It is also the location of the pineal gland – the source of melatonin, a hormone central to the sleep–wake cycle and body clock.

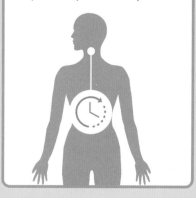

RESPONSES OF THE HYPOTHALAMUS

STIMULUS	RESPONSE
Day length	Helps to maintain body rhythms after receiving signals about day length from the optical system.
Water	When the blood's water levels drop, releases vasopressin, also called antidiuretic hormone, which reduces the volume of urine.
Eating	When the stomach is full, releases leptin to reduce feelings of hunger.
Lack of food	When the stomach is empty, releases ghrelin to boost feelings of hunger.
Infection	Increases body temperature to help the immune system work faster to fight off pathogens.
Stress	Increases the production of cortisol, a hormone associated with preparing the body for a period of physical activity.
Body activity	Stimulates the production of thyroid hormones to boost the metabolism, and somatostatin to reduce it.
Sexual activity	Organizes the release of oxytocin, which helps the formation of interpersonal bonds. The same hormone is released during childbirth.

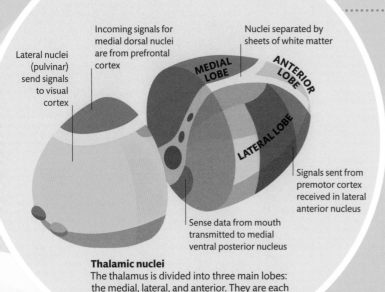

Lateral nuclei (pulvinar) send signals to visual cortex

Incoming signals for medial dorsal nuclei are from prefrontal cortex

Nuclei separated by sheets of white matter

MEDIAL LOBE

ANTERIOR LOBE

LATERAL LOBE

Signals sent from premotor cortex received in lateral anterior nucleus

Sense data from mouth transmitted to medial ventral posterior nucleus

Thalamic nuclei
The thalamus is divided into three main lobes: the medial, lateral, and anterior. They are each further organized into zones, or nuclei, associated with particular sets of functions.

The thalamus

The word thalamus is derived from the Greek word for "inner chamber", and this thumb-sized mass of grey matter sits in the middle of the brain, between the cerebral cortex and midbrain. It is formed from several bundles, or tracts, of nerves, which send and receive signals in both directions between the upper and lower regions of the brain, often in feedback loops (see p.91). It is associated with the control of sleep, alertness, and consciousness. Signals from every sensory system, except smell, are directed through the thalamus to the cortex for processing.

WEIGHING JUST **4 G (0.1 OZ)**, THE **HYPOTHALAMUS** IS NOT MUCH LARGER THAN THE END SEGMENT OF **A LITTLE FINGER**

The pituitary gland

Weighing about 0.5 g (0.01 oz), the tiny pituitary gland produces many of the body's most significant hormones under the direction of the hypothalamus. The hormones are released into the blood supply via a network of tiny capillaries. Pituitary hormones include those that control growth, urination, the menstrual cycle, childbirth, and skin tanning. Despite having the volume of a pea, the gland is divided into two main lobes, the anterior and posterior, plus a small intermediate lobe. Each lobe is devoted to the production of a particular set of hormones.

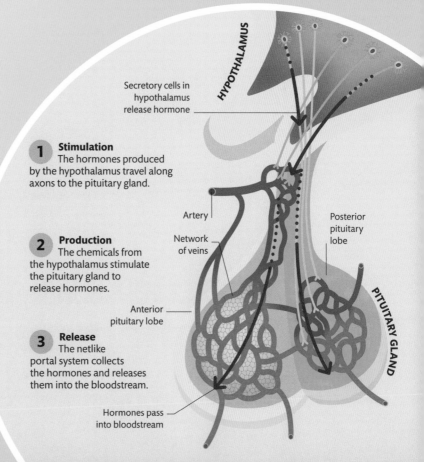

HYPOTHALAMUS

Secretory cells in hypothalamus release hormone

1 Stimulation
The hormones produced by the hypothalamus travel along axons to the pituitary gland.

2 Production
The chemicals from the hypothalamus stimulate the pituitary gland to release hormones.

3 Release
The netlike portal system collects the hormones and releases them into the bloodstream.

Artery

Network of veins

Anterior pituitary lobe

Posterior pituitary lobe

PITUITARY GLAND

Hormones pass into bloodstream

Thalamus links brainstem
with forebrain, relaying
and pre-processing
sensory and other
information

THALAMUS

Midbrain is associated with
control of state of arousal
and body temperature

MIDBRAIN

HOW BIG IS THE CEREBELLUM?

Most of the brain's cells are located in the cerebellum, although it makes up only around 10 per cent of the volume of the whole brain.

Connecting the brain

The stalklike brainstem forms a link between the thalamus, the base of the forebrain, and the spinal cord, which connects to the rest of the body. It is involved in many basic functions, including the sleep-wake cycle, eating, and regulating heart rate.

Pons is a major
communication pathway
that carries cranial nerves
used for breathing, hearing,
and eye movements

The brainstem

The brainstem is made up of three components, all of which have an essential role in several of the human body's most fundamental functions. The midbrain is the start point of the reticular formation, a series of brain nuclei (see pp.32–33) that run through the brainstem and are linked to arousal and alertness and play a crucial role in consciousness. The pons is another series of nuclei that send and receive signals from the cranial nerves associated with the face, ears, and eyes. The medulla descends and narrows to merge with the uppermost end of the spinal cord. It handles many of the autonomous body functions, such as blood-pressure regulation, blushing, and vomiting.

BRAINSTEM

PONS

MEDULLA

THALAMUS

BRAINSTEM

CEREBELLUM

10 pairs of
cranial nerves
emerge from
brainstem

Cranial nerves
start and end
at nuclei in
brainstem

Medulla is involved in
important reflexes
such as breathing rate
and swallowing

The brainstem and cerebellum

The lower regions of the brain are the brainstem, which connects directly to the spinal cord, and the cerebellum, located directly behind it.

Spinal cord consists
of a bundle of nerve
axons that connect
to peripheral
nervous system

REAR VIEW OF CEREBELLUM

The little brain

The cerebellum, a term that means "little brain", is a highly folded region of the hindbrain that sits behind the brainstem. Like the cerebrum above it, the cerebellum is divided into two lobes. These are divided laterally into functional zones.

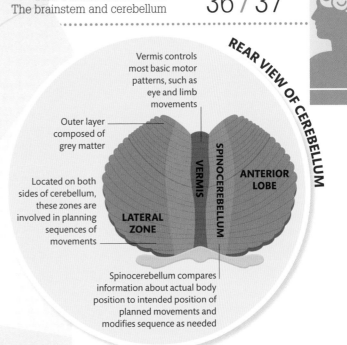

Vermis controls most basic motor patterns, such as eye and limb movements

Outer layer composed of grey matter

Located on both sides of cerebellum, these zones are involved in planning sequences of movements

SPINOCEREBELLUM

VERMIS

ANTERIOR LOBE

LATERAL ZONE

Spinocerebellum compares information about actual body position to intended position of planned movements and modifies sequence as needed

Anterior lobe of cerebellum receives information about body posture from spinal cord

Body movements are coordinated in posterior lobe

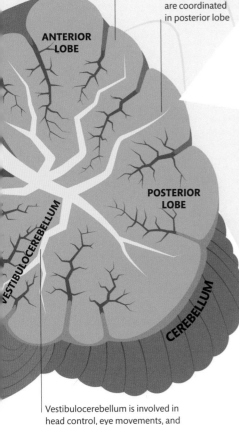

ANTERIOR LOBE

POSTERIOR LOBE

VESTIBULOCEREBELLUM

CEREBELLUM

Vestibulocerebellum is involved in head control, eye movements, and maintaining balance through information from the inner ear

The cerebellum

Although the cerebellum appears to play a part in maintaining attention and processing language, it is most associated with its role in the regulation of body movement. Specifically, its role is to convert the broad executive motor requests into smooth and coordinated muscle sequences, error-correcting all the while. It routes its outputs through the thalamus. At the microscopic level, the cerebellum's cells are arranged in layers. The purpose of these layers is to lay down fixed neural pathways for all kinds of learned movement patterns, such as walking, talking, and keeping balance. Damage to the cerebellum does not result in paralysis, but slow jerky movements.

KNOWLEDGE OF THE CEREBELLUM WAS ADVANCED BY STUDYING **BRAIN-INJURED SOLDIERS** IN **WWI**

THE CEREBELLUM AND NEURAL NETWORKS

Some artificial intelligences (AI) use a system inspired by the anatomy of the cerebellum. AI programs itself by machine learning. It does this with a processor called a neural network, where inputs find their way by trial and error through layers of connections, a set-up that mirrors the way the cerebellum lays down patterns for learned movements.

The limbic system

Sitting below the cortex and above the brainstem, the limbic system is a collection of structures associated with emotion, memory, and basic instincts.

Location and function

The limbic system is a cluster of organs situated in the centre of the brain, occupying parts of the medial surfaces of the cerebral cortex. Its major structures form a group of modules that pass signals between the cortex and the bodies of the lower brain. Nerve axons link all of its parts and connect them to other brain areas. The limbic system mediates instinctive drives such as aggression, fear, and appetite, with learning, memory, and higher mental activities.

System parts

The limbic system's components extend from the cerebrum inwards and down to the brainstem. It is usually understood to include the structures shown here.

THE S-SHAPED **HIPPOCAMPUS** IS NAMED AFTER ITS RESEMBLANCE TO A **SEAHORSE**

Fornix is a bundle of nerve tracts that connects hippocampus to thalamus and lower brain beneath

CINGULATE GYRUS

COLUMN OF FORNIX

FORNIX

MAMILLARY BODIES

MIDBRAIN

HYPOTHALAMUS

AMYGDALA

PARAHIPPOCAMPAL GYRUS

OLFACTORY BULB

SENSE OF SMELL

Smell, which is processed in the olfactory bulbs, is the only sense handled by the limbic system and not sent through the thalamus.

NEW MEMORIES

The small mamillary bodies act as relay stations for new memories formed in the hypothalamus. Damage leads to an inability to sense direction, particularly with regards to location.

FEAR CONDITIONING

The amygdala is most associated with fear conditioning, where we learn to be afraid of something. It is also involved in memory and emotional responses.

RECOGNITION

Involved in forming and retrieving memories associated with fresh data from the senses, the parahippocampal gyrus helps us recognize and recall things.

WHAT DOES LIMBIC MEAN?

The word limbic is derived from the Latin limbus, meaning "border", referring to the system's role as a kind of transition zone between the cortex and lower brain.

Reward and punishment

The limbic system is closely linked to feelings of rage and contentment. Both are due to the stimulation of reward or punishment centres within the limbic system, particularly in the hypothalamus. Reward and punishment are crucial aspects of learning, in that they create a basic response to experiences. Without this rating system, the brain would simply ignore old sensory stimuli that it had already experienced and only pay attention to new stimuli.

Pleasure
Associated with the release of dopamine, the brain seeks to repeat behaviours that create this feeling.

Disgust
This emotion is linked to the sense of smell. Its primordial role is to protect us from infection.

Fear
Fear is linked to specific stimuli by the amygdala. This can lead to a controlled rage or fight response.

Cingulate gyrus helps to form memories associated with strong emotion

HIPPOCAMPUS

EPISODIC MEMORIES

The hippocampus receives and processes inputs from the cerebrum. It is involved in creating episodic memories, or memories about what you have done, and creating spatial awareness.

Klüver-Bucy syndrome

This condition is caused by damage to the limbic system and results in a spectrum of symptoms associated with the loss of fear and impulse control. First described in humans in 1975, this neural disorder is named after the 1930s investigators Heinrich Klüver and Paul Bucy, who performed experiments that involved removing various brain regions in live monkeys and noting the effects.

In humans, the syndrome may be caused by Alzheimer's disease, complications from herpes, or brain damage. It was first documented in people who had undergone surgical removal of parts of the brain's temporal lobe. The condition can be treated with medication and assistance with daily tasks.

SYMPTOM	DESCRIPTION
Amnesia	Damage to the hippocampus leads to the inability to form long-term memories.
Docility	With little sensation of reward for actions, sufferers lack motivation.
Hyperorality	An urge to examine objects by putting them in the mouth.
Pica	Eating compulsively, including inedible substances like earth.
Hypersexuality	A high sex drive often associated with fetishes or atypical attractions.
Agnosia	Losing the ability to recognize familiar objects or people.

Imaging the brain

Modern medicine and neuroscience can see through the skull to observe structures within the living brain. However, imaging this soft and intricate organ has required the invention of advanced technology.

MRI scanners

A magnetic resonance imaging (MRI) machine gives the best general view of the brain's nervous tissue and is most often deployed to search for tumours. MRI does not expose the brain to high-energy radiation, unlike other scanning systems, which makes it safe to use for long periods and multiples times. Two refinements of MRI, called fMRI and DTI, are also useful for monitoring brain activity (see p.43). Although ideal as a tool for research and diagnosis, MRI is expensive. With its liquid-helium coolant system and superconducting electromagnets, one machine also uses the power of six family homes.

How MRI works

MRI makes use of the way that protons in hydrogen atoms align to magnetic fields. Hydrogen is found in water and fats, which are both common in the brain. A scan takes about an hour, then the data is processed to create detailed images.

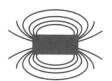

THE ELECTROMAGNET IN AN MRI SCANNER CAN GENERATE A MAGNETIC FIELD 40,000 TIMES AS STRONG AS EARTH'S

Layer of thermal insulation keeps liquid helium cold

Liquid helium cools electromagnet to about -270°C (-453°F)

Superconducting electromagnet generates extremely strong magnetic field

Gradient magnets focus magnetic field around area to be scanned

Radiofrequency coil emits and detects radio waves

Patient lies inside body of scanner during scanning

Motorized table moves patient into scanner

LIQUID HELIUM

MOTORIZED TABLE

Protons aligned randomly

INACTIVE ELECTROMAGNET

INACTIVE ELECTROMAGNET

Additional south-facing proton

ACTIVE ELECTROMAGNET

Proton faces south

Magnetic field line

ACTIVE ELECTROMAGNET

Proton faces north

1 **Protons unaligned**
Before the MRI machine is activated, the protons in the brain's molecules are unaligned – the axes around which the particles are spinning point in random directions.

2 **Protons align to magnetic field**
Activating the machine's powerful magnetic field forces all the protons to align with each other. Approximately half face the field's north pole, and half face south. However, one pole will always have slightly more protons facing it than the other.

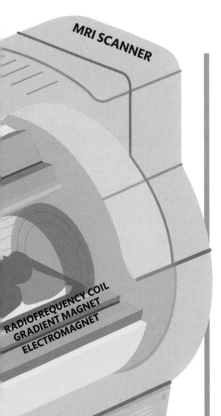

MRI SCANNER

RADIOFREQUENCY COIL
GRADIENT MAGNET
ELECTROMAGNET

CT scans

Computer tomography (CT), or computerized axial tomography (CAT), takes a series of X-ray images through the brain from different angles. A computer then compares the images to create a single cross-section of the brain. CT scans are quicker than MRI and are best for detecting strokes, skull fractures, and brain haemorrhages.

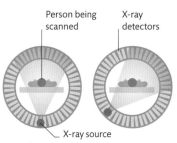

Person being scanned

X-ray detectors

X-ray source

Rotating X-ray
The X-ray source shines through the brain, arcing around the patient to vary the angle of each image.

OTHER TYPES OF SCANNING TECHNOLOGY	
Imaging certain brain features require particular scanning techniques, which may also be used if MRI or CT are dangerous or unsuitable.	
TYPE OF SCAN	**TECHNOLOGY AND USES**
PET (positron emission tomography)	Used in order to image the blood flow through the brain and highlight active regions. PET scans track the location of radioactive tracers injected into the blood.
DOI (diffuse optical imaging)	An array of newer techniques that works by detecting how bright light or infrared rays penetrate into the brain. DOI provides a way of monitoring blood flow and brain activity.
Cranial ultrasound	A safe imaging technique that is based on the way ultrasonic waves bounce off structures in the brain. Cranial ultrasound is mostly used on infants. It is used less often in adults because the images lack detail.

RADIOFREQUENCY COIL

Radio-wave pulse

RADIOFREQUENCY COIL

Additional proton flips into different orientation

Flipped proton realigns

RADIOFREQUENCY COIL

RADIOFREQUENCY COIL

Radio signal emitted

Computer processes signal data

Image shows tissues in cross-section

COMPUTER

MONITOR

Radiofrequency coil detects signal and passes it to computer

3 A pulse of radio waves
With the magnetic field on, the MRI machine's radiofrequency coil sends a pulse of radio waves through the brain. This input of extra energy makes the spare protons flip out of alignment.

4 Radio signal emitted
Once the pulse is switched off, the unaligned protons flip back into alignment with the magnetic field. This causes them to release energy as a radio signal, which is detected by the machine.

5 Receiver creates image
All the signal data is then processed by computer to create two-dimensional "slices" of the brain. Protons in different body tissues produce different signals, so scans can show the tissues distinctly and in great detail.

Monitoring the brain

Being able to collect information from a living brain at work has revolutionized both our understanding of how the brain functions and brain medicine.

EEG

The simplest brain monitor is the electroencephalograph (EEG). It uses electrodes positioned all over the cranium to pick up an electrical field created by the activity of neurons in the cerebral cortex. The varying levels may be displayed as waves ("ordinary EEG") or coloured areas (quantitative EEG, or QEEG). EEG can reveal evidence of seizure disorders, such as epilepsy, and signs of injury, inflammation, and tumours. The painless procedure is also used to assess brain activity in coma patients.

WHY DOES THE BRAIN PRODUCE ELECTROMAGNETIC FIELDS?

Neurons use pulses of electric charge to transmit messages. The activity of billions of cells accumulates into a constant field.

Types of EEG wave
Neighbouring cells in the cortex fire in synchrony, creating wavelike changes in the intensity of the electrical field. Characteristic wave patterns (named after letters of the Greek alphabet) have been found to be closely associated with certain brain states.

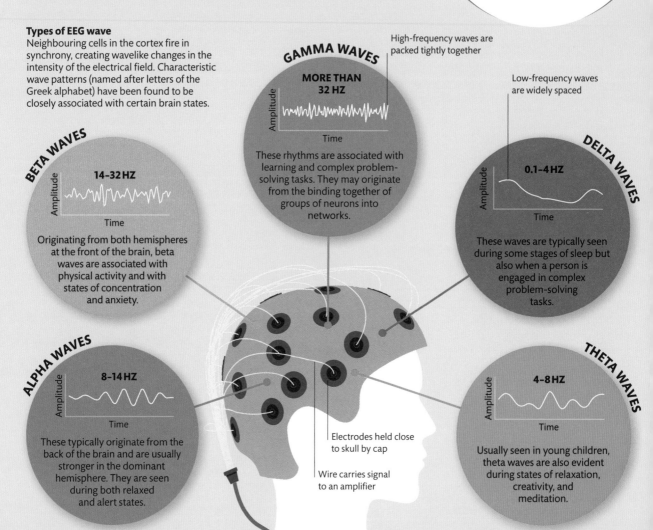

GAMMA WAVES
MORE THAN 32 HZ
Amplitude / Time
High-frequency waves are packed tightly together
These rhythms are associated with learning and complex problem-solving tasks. They may originate from the binding together of groups of neurons into networks.

BETA WAVES
14–32 HZ
Amplitude / Time
Originating from both hemispheres at the front of the brain, beta waves are associated with physical activity and with states of concentration and anxiety.

DELTA WAVES
0.1–4 HZ
Amplitude / Time
Low-frequency waves are widely spaced
These waves are typically seen during some stages of sleep but also when a person is engaged in complex problem-solving tasks.

ALPHA WAVES
8–14 HZ
Amplitude / Time
These typically originate from the back of the brain and are usually stronger in the dominant hemisphere. They are seen during both relaxed and alert states.

THETA WAVES
4–8 HZ
Amplitude / Time
Usually seen in young children, theta waves are also evident during states of relaxation, creativity, and meditation.

Electrodes held close to skull by cap

Wire carries signal to an amplifier

MEG

As well as making electrical activity, the brain produces a faint magnetic field. This is detected by a magnetoencephalography (MEG) machine and can be used to create a real-time account of activity in the cerebral cortex. MEG is limited by the weakness of the brain's magnetism, but the technique can detect rapid fluctuations in brain activity, which take place over a few thousandths of a second, better than other monitoring systems.

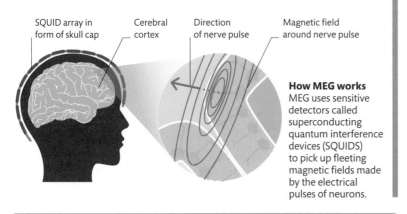

SQUID array in form of skull cap

Cerebral cortex

Direction of nerve pulse

Magnetic field around nerve pulse

How MEG works
MEG uses sensitive detectors called superconducting quantum interference devices (SQUIDS) to pick up fleeting magnetic fields made by the electrical pulses of neurons.

Functional MRI and diffusion tensor imaging

MRI (see pp.40–41) can be extended to collect information about what the brain is doing. Functional MRI (fMRI) scanning tracks the flow of blood through the brain, specifically showing where it is giving oxygen to neurons and thus indicating which regions are active in real time. Subjects are asked to carry out mental and physical tasks while monitored by fMRI to create a functional map of the brain and spinal cord that combines anatomy with activity levels. Diffuse tensor imaging (DTI) also uses MRI but tracks the natural movement of water through brain cells. It is used to build up a map of the white-matter connections within the brain.

NEUROFEEDBACK

This form of cognitive therapy uses an EEG to create a feedback loop between a person's mental state and their brain activity. This makes it easier for people to learn ways to control unwanted mental activity, such as anxiety.

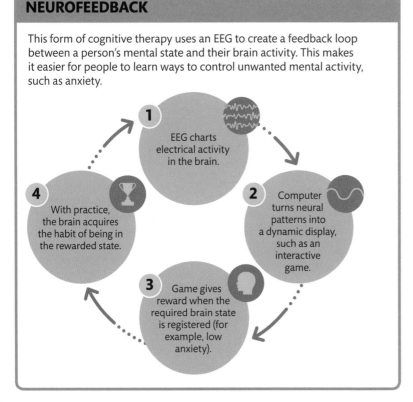

1 EEG charts electrical activity in the brain.

2 Computer turns neural patterns into a dynamic display, such as an interactive game.

3 Game gives reward when the required brain state is registered (for example, low anxiety).

4 With practice, the brain acquires the habit of being in the rewarded state.

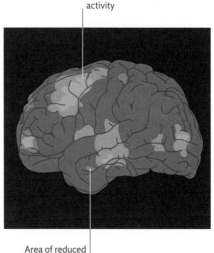

Area of increased activity

Area of reduced activity

Interpreting an fMRI image
An fMRI scan begins with establishing a baseline of activity in the brain. The scan then shows up regions that fluctuate from this baseline, allowing researchers to work out which areas are excited or inhibited during particular tasks.

Brain development

The first nerve cells are produced just days after conception. These cells form into a plate and then curl to become a liquid-filled structure, called the neural tube, which develops into the brain and spinal cord. One end becomes a bulge and then splits into distinct areas.

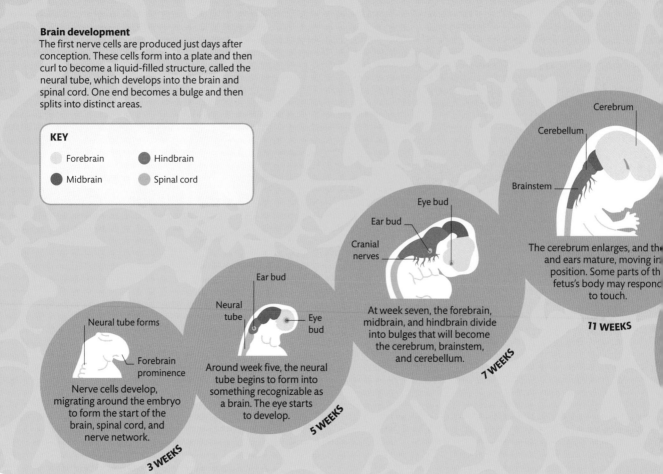

KEY

- Forebrain
- Midbrain
- Hindbrain
- Spinal cord

Neural tube forms

Forebrain prominence

Nerve cells develop, migrating around the embryo to form the start of the brain, spinal cord, and nerve network.

3 WEEKS

Neural tube

Ear bud

Eye bud

Around week five, the neural tube begins to form into something recognizable as a brain. The eye starts to develop.

5 WEEKS

Eye bud

Ear bud

Cranial nerves

At week seven, the forebrain, midbrain, and hindbrain divide into bulges that will become the cerebrum, brainstem, and cerebellum.

7 WEEKS

Cerebrum

Cerebellum

Brainstem

The cerebrum enlarges, and the and ears mature, moving ir. position. Some parts of th fetus's body may respond to touch.

11 WEEKS

Babies and young children

The human brain begins to develop after conception and changes rapidly for the first few years of life, but it takes more than 20 years for a brain to fully mature.

Before birth

An embryo's brain has a lot of development to do, going from just a few nerve cells three weeks after conception to an organ with specialized areas that is ready to start learning from birth. Genes control this process, but the environment can affect it as well. Insufficient nutrition can change brain development, and extreme stress on the mother during pregnancy can have an impact, too.

RECOGNIZING FACES

Babies prefer looking at facelike images, and learn about faces rapidly. An area of the cortex called the face recognition area (see p.68) becomes specialized in identifying faces. Chess champions also use this area to recognize board layouts, suggesting that the most important patterns in a person's life are decoded there.

FACELIKE

NOT FACELIKE

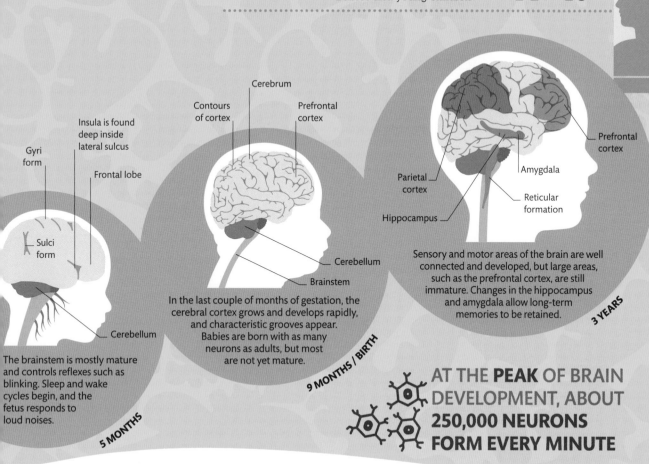

Cerebrum

Contours of cortex

Prefrontal cortex

Insula is found deep inside lateral sulcus

Frontal lobe

Gyri form

Prefrontal cortex

Parietal cortex

Amygdala

Reticular formation

Hippocampus

Sulci form

Cerebellum

Brainstem

Cerebellum

Sensory and motor areas of the brain are well connected and developed, but large areas, such as the prefrontal cortex, are still immature. Changes in the hippocampus and amygdala allow long-term memories to be retained.

3 YEARS

In the last couple of months of gestation, the cerebral cortex grows and develops rapidly, and characteristic grooves appear. Babies are born with as many neurons as adults, but most are not yet mature.

9 MONTHS / BIRTH

The brainstem is mostly mature and controls reflexes such as blinking. Sleep and wake cycles begin, and the fetus responds to loud noises.

5 MONTHS

AT THE **PEAK** OF BRAIN DEVELOPMENT, ABOUT **250,000 NEURONS** FORM EVERY MINUTE

Children's brains

After birth, babies' brains are like sponges; they are incredible at taking in information from the world around them and trying to make sense of it. During the first few years, the brain grows and develops rapidly, with brain volume doubling in the first year of life. Synapses grow and form new connections quickly and easily, a process called neuroplasticity.

Building connections
Peak plasticity for each region of the brain is different. Sensory areas build synapses rapidly four to eight months after birth, but prefrontal areas do not reach peak plasticity until an infant is around 15 months old.

NEWBORN

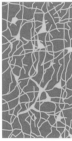
9 MONTHS

2 YEARS

WHY IS OUR BRAIN WRINKLY?

As human intelligence evolved, our cortex expanded. But bigger heads would mean that babies could not fit through the birth canal. A folded cortex packs more tissue into a smaller volume.

Older children and teenagers

Teenage brains undergo dramatic restructuring. Unused connections are pruned, and insulating myelin coats the most important connections, making them more efficient.

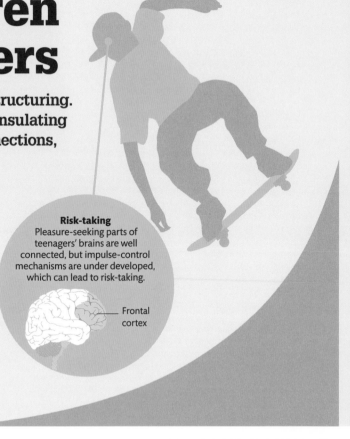

Teenage behaviour

Teenagers have a reputation for being impulsive, rebellious, self-centred, and emotional. A lot of this is due to the changes happening in adolescent brains. Human brains change and develop in set patterns, leaving teenagers with a mix of mature and immature brain regions as they grow. The last area to fully develop is the frontal cortex, which regulates the brain and controls impulses. This area allows adults to exert self-control over their emotions and desires, which is something adolescents can struggle with.

Risk-taking
Pleasure-seeking parts of teenagers' brains are well connected, but impulse-control mechanisms are under developed, which can lead to risk-taking.

Frontal cortex

Sleep cycles

During our teenage years, we need plenty of sleep as our brain continues to develop. But at this time, our circadian rhythms shift as melatonin, the hormone which is released in the evening and makes us feel sleepy, begins to be released later than usual. This is why teenagers often want to go to bed later than children and adults, and may struggle to get up for school in the morning.

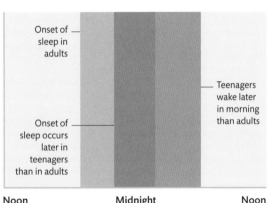

Onset of sleep in adults

Teenagers wake later in morning than adults

Onset of sleep occurs later in teenagers than in adults

Noon · Midnight · Noon

KEY
- Adult sleep time
- Adolescent sleep time

Out of sync
Waking teenagers early for school is like giving them constant jet lag. Studies have shown that starting school an hour later improved attendance and grades. Fights and even car accidents also decreased.

SYNAPTIC PRUNING

Synaptic pruning, which is when unused neural connections die off, starts during childhood and continues through our teen years. Cortical areas are pruned from the back to the front. Pruning makes each area more efficient, so until it is finished, that region cannot be considered fully mature.

IMMATURE

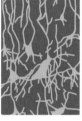

MATURE

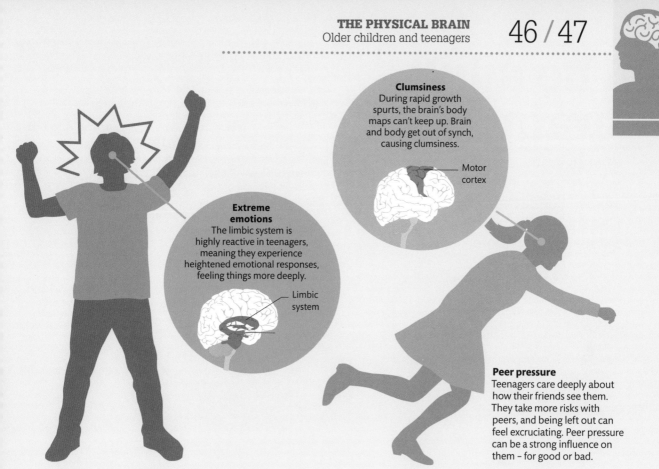

Clumsiness
During rapid growth spurts, the brain's body maps can't keep up. Brain and body get out of synch, causing clumsiness.

Motor cortex

Extreme emotions
The limbic system is highly reactive in teenagers, meaning they experience heightened emotional responses, feeling things more deeply.

Limbic system

Peer pressure
Teenagers care deeply about how their friends see them. They take more risks with peers, and being left out can feel excruciating. Peer pressure can be a strong influence on them – for good or bad.

Mental-health risks

Some of the brain areas that undergo the most dramatic changes during adolescence have been linked with mental ill-health. These changes can leave the brain vulnerable to small issues becoming dysfunctions. This may explain why so many mental health problems, from schizophrenia to anxiety disorders, commonly appear during adolescence.

THE BRAIN REACHES ITS
LARGEST PHYSICAL SIZE BETWEEN
AGES 11 AND 14

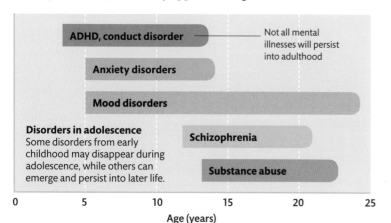

ADHD, conduct disorder — Not all mental illnesses will persist into adulthood

Anxiety disorders

Mood disorders

Disorders in adolescence
Some disorders from early childhood may disappear during adolescence, while others can emerge and persist into later life.

Schizophrenia

Substance abuse

| 0 | 5 | 10 | 15 | 20 | 25 |

Age (years)

WHY ARE TEENS SELF-CONSCIOUS?

When we think about being embarrassed, a region of our prefrontal cortex linked to understanding mental states is more active in teenagers than adults.

The adult brain

Human brains continue to change and mature throughout early adulthood, as unused connections are pruned away. This makes the brain more efficient but also less flexible.

PARENTHOOD

A new mother's brain and body are awash with hormones such as oxytocin, driving her to care for her baby. Looking at her infant triggers the brain's reward pathways, and her amygdala becomes more active, scanning for danger. Men's brains are affected by parenthood too, but only if they spend a lot of time with their baby. The brains of men who are primary caregivers of an infant go through similar changes to women's, and these changes appear to be very similar to falling in love.

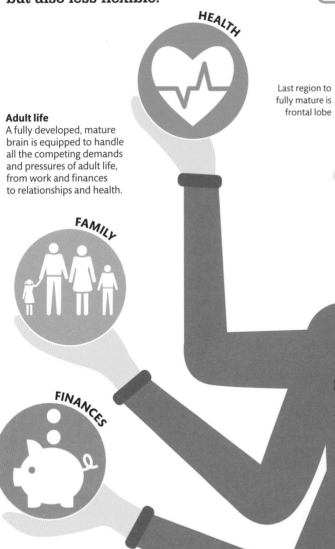

HEALTH

Adult life
A fully developed, mature brain is equipped to handle all the competing demands and pressures of adult life, from work and finances to relationships and health.

FAMILY

FINANCES

Corpus callosum is fully developed to allow information flow between hemispheres

Last region to fully mature is frontal lobe

Amygdala is less emotionally reactive

Hippocampus continues to produce new brain cells

Mature brains
Full myelination (the sheathing of axons in myelin) allows information to flow freely, but the process is only completed in a person's late 20s. The last brain region to finish maturing is the frontal lobe, which is responsible for judgment and inhibition. Compared to children and teenagers, adults are better able to regulate their emotions and control their impulses. They can use their experiences to better predict the outcomes of their actions and how they may make other people feel.

THE VOLUME OF **WHITE MATTER** IN A PERSON'S BRAIN **PEAKS AROUND AGE 40**

MORALITY

FUTURE

WORK

Neurogenesis

Neurogenesis is the development of new neurons by neural stem cells (cells that can become other cells). In a range of mammals, neurogenesis happens in the hippocampus and olfactory areas, and continues throughout life, with new neurons being produced regularly.
The same is thought to be true in humans, although the evidence is mixed. Neurogenesis may also play a role in learning and memory.

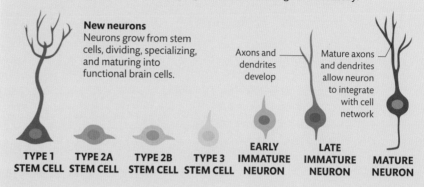

New neurons
Neurons grow from stem cells, dividing, specializing, and maturing into functional brain cells.

Axons and dendrites develop

Mature axons and dendrites allow neuron to integrate with cell network

TYPE 1 STEM CELL | TYPE 2A STEM CELL | TYPE 2B STEM CELL | TYPE 3 STEM CELL | EARLY IMMATURE NEURON | LATE IMMATURE NEURON | MATURE NEURON

Disrupting memories

New brain cells help store information, so boosting neurogenesis in the brain can improve learning into adulthood. However, it also has a role to play in forgetting. Adding in new brain cells with new connections disrupts existing memory circuits by competing with them. This means there is an optimal level of neurogenesis, which balances learning ability with retaining older memories.

Memory storage
Due to the creation of new brain cells, hippocampal memories may degrade before they can be stored in the cortex. This might explain why we are unable to remember our infancy.

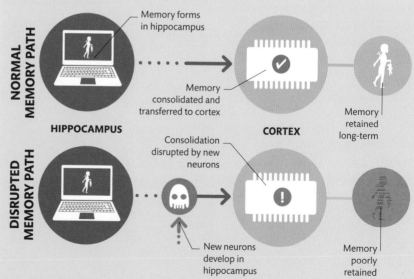

NORMAL MEMORY PATH

Memory forms in hippocampus

Memory consolidated and transferred to cortex

HIPPOCAMPUS

CORTEX

Memory retained long-term

DISRUPTED MEMORY PATH

Consolidation disrupted by new neurons

New neurons develop in hippocampus

Memory poorly retained

The ageing brain

With age, some abilities decline as neurons degenerate and the brain decreases in volume. In those neurons that remain, impulses may travel more slowly.

The shrinking brain

As we age, there is a natural reduction of neurons as they degenerate, and the brain as a whole shrinks 5 to 10 per cent in volume. This is partially due to decreased blood flow to ageing brains. The fatty myelin that insulates the axons of neurons also decays with age, leaving brain circuits less efficient at transmitting information, which can lead to problems with memory recall and maintaining balance.

KEY

- ● Grey matter
- ○ White matter
- ● Basal ganglia
- ● Ventricles

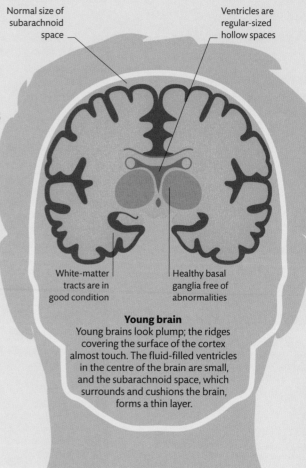

Normal size of subarachnoid space

Ventricles are regular-sized hollow spaces

White-matter tracts are in good condition

Healthy basal ganglia free of abnormalities

Young brain
Young brains look plump; the ridges covering the surface of the cortex almost touch. The fluid-filled ventricles in the centre of the brain are small, and the subarachnoid space, which surrounds and cushions the brain, forms a thin layer.

Ageing and happiness

Ageing might seem like a bad thing, but studies have shown that as we get older our feelings of happiness and well-being increase, while levels of stress and worry decrease. Older adults' brains seem to be better at focussing on the positive. They are more likely to remember happy than sad pictures, and spend more time looking at happy faces than angry or upset ones.

Ups and downs
A study found younger and older people reported higher levels of well-being than those in middle age. Happiness levels rose steadily from age 50 onwards.

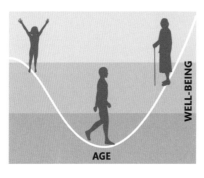

WELL-BEING

AGE

ALZHEIMER'S DISEASE

Alzheimer's disease, the most common form of dementia (see p.200), is linked to the build-up of proteins in the brain, which clump into plaques and tangles. Eventually, affected brain cells die, causing memory loss and other symptoms. Scientists do not know yet whether the proteins cause the disease or are a symptom of it, and drugs to break them down have not helped patients.

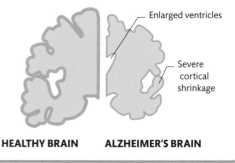

Enlarged ventricles

Severe cortical shrinkage

HEALTHY BRAIN   **ALZHEIMER'S BRAIN**

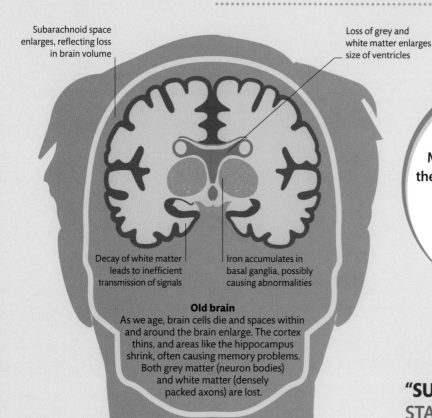

Subarachnoid space enlarges, reflecting loss in brain volume

Loss of grey and white matter enlarges size of ventricles

Decay of white matter leads to inefficient transmission of signals

Iron accumulates in basal ganglia, possibly causing abnormalities

Old brain
As we age, brain cells die and spaces within and around the brain enlarge. The cortex thins, and areas like the hippocampus shrink, often causing memory problems. Both grey matter (neuron bodies) and white matter (densely packed axons) are lost.

CAN WE TREAT ALZHEIMER'S?

Medication can slow down the progression of the disease, and manage some of the symptoms, but a cure for Alzheimer's has not yet been found.

"SUPER-AGERS'" BRAINS STAY LOOKING YOUNG FOR THEIR WHOLE LIVES

A slow decline?

As we get older, our attention suffers, and our brains become less plastic. This makes learning harder, although not impossible. In fact, learning new things throughout life boosts brain health and may stave off cognitive decline by strengthening neural synapses. And with age come some benefits: on average, older adults are better at extracting the big picture from a situation and using their life experience to solve problems.

Skills and abilities
The Seattle Longitudinal Study followed adults for 50 years. It found that skills like vocabulary and general knowledge keep improving for most of our lives.

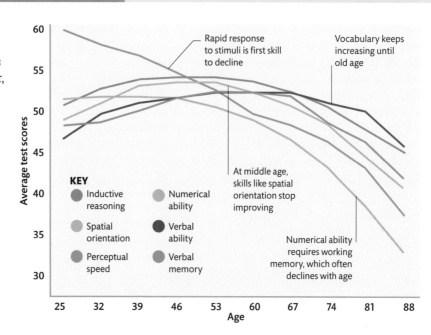

As we get older, most of us notice a slight reduction in thinking speed as well as a reduction in our working memory (see p.135). Some people experience severe decline or even dementia (see p.200), but this is by no means inevitable. In fact, some cognitive capacities, such as our overall understanding of life, may even improve as we age.

We inherit our basic level of cognitive function from our parents, but this genetic blueprint is also affected by our environment and life experiences, including nutrition, health, education, stress levels, and relationships. Physically, socially, and intellectually stimulating activities also play a key role.

Preventing decline

We can take a variety of steps to safeguard our brain's health. A diet high in vegetables, fruit, "good" fats, and nutrients (see pp.54–55) keeps both brain and body healthy, as does moderate but regular physical activity. Jogging or other aerobic exercise can help delay age-related declines both in memory and thinking speed.

You can also protect your brain health by avoiding toxins, such as alcohol and tobacco. Smoking has been linked with damage to the brain's cortex. If you do drink alcohol, keep within healthy drinking limits, and have at least two alcohol-free days per week.

Keep your mind stimulated. Any mental challenge that involves learning – from home repairs to cooking to crossword puzzles – can stretch cognitive skills. Consider learning a new language, as people who speak two or more languages have stronger cognitive ability than those who speak just one.

To sum up, you can slow the cognitive ageing process by:
- **Keeping your brain well supplied with oxygen and nutrients.**
- **Avoiding exposure to toxins such as alcohol and nicotine.**
- **Exercising your body by building activity into daily life.**
- **Exercising your mind by learning new skills.**

How to slow the effects of ageing

As we age, our thinking and short-term memory may become less efficient. Nevertheless, we continue to learn until we die, and we can take active measures to keep our brain working well at any age.

Brain food

Like any other organ, the human brain needs a constant supply of water and nutrients to maintain its health and to supply energy for efficient functioning.

Feeding the brain

A healthy diet benefits both the mind and the body. Complex carbohydrates provide a steady flow of fuel; these are found in wholegrain bread, brown rice, legumes, potatoes, and sweet potatoes. Healthy fats are essential for maintaining brain cells, and these fats come from oily fish, vegetable oils, and plant foods such as avocados and flax seeds. Proteins supply amino acids. Fruits and vegetables supply water, vitamins, and fibre.

HYDRATION

Brain cells need adequate hydration (water supply) in order to function effectively. Studies have shown that dehydration can impair our ability to concentrate and to perform mental tasks, and negatively affect memory. Some of our water intake comes from the food we consume, but it is helpful to drink several glasses of water each day to maintain a healthy level of hydration.

Sources of nutrients
Fresh fruits and vegetables, beans and lentils, whole grains, healthy fats such as olive oil, and oily fish such as salmon, all supply vital nutrients for the brain.

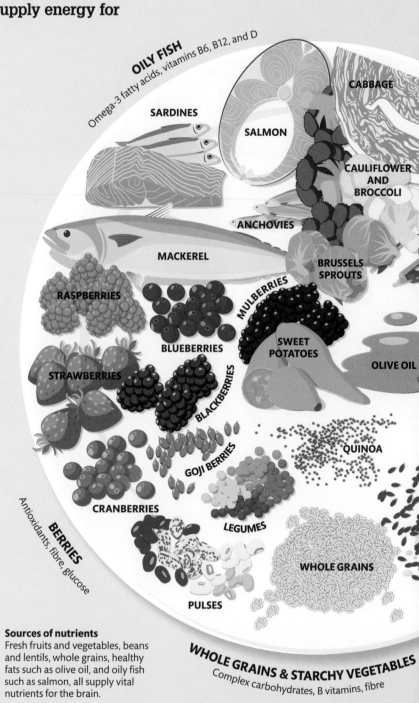

OILY FISH
Omega-3 fatty acids, vitamins B6, B12, and D

SARDINES

SALMON

CABBAGE

CAULIFLOWER AND BROCCOLI

ANCHOVIES

MACKEREL

BRUSSELS SPROUTS

RASPBERRIES

MULBERRIES

SWEET POTATOES

OLIVE OIL

BLUEBERRIES

STRAWBERRIES

BLACKBERRIES

QUINOA

GOJI BERRIES

CRANBERRIES

LEGUMES

WHOLE GRAINS

BERRIES
Antioxidants, fibre, glucose

PULSES

WHOLE GRAINS & STARCHY VEGETABLES
Complex carbohydrates, B vitamins, fibre

THE **BRAIN IS 60 PER CENT FAT** AND NEEDS A **STEADY SUPPLY OF ENERGY**

CRUCIFEROUS VEGETABLES & DARK LEAFY GREENS
Antioxidants, fibre, nutrients

KALE

SPINACH

CHARD

OLIVES

VEGETABLE OIL

OLIVE & VEGETABLE OILS
Omega-3 and omega-6, monounsaturated fats

FLAX SEED / OIL

Essential nutrients

Certain nutrients from food have been found to improve or maintain particular brain functions. These substances include vitamins and minerals, omega-3 and omega-6 fatty acids, antioxidants, and water. These essential nutrients help to keep brain cells healthy, enable the cells to transmit signals quickly and effectively, reduce damage from inflammation and free radicals (atoms that can damage cells, proteins, and DNA), and help the cells to form new connections. They can also promote the production and function of neurotransmitters. As a result, regularly eating foods that contain these nutrients can benefit memory, cognitive functions, concentration, and mood.

NUTRIENT	BENEFIT	SOURCE
Omega-3 and omega-6 fatty acids	Help maintain blood flow and cell membranes in brain; support memory and reduce risk of depression, mood disorders, stroke, and dementia	Oily fish (such as salmon, sardines, herring, mackerel) Flaxseed oil, rapeseed oil Walnuts, pine nuts, Brazil nuts
B vitamins	Vitamins B6, B12, and folic acid, support nervous-system function; choline helps production of neurotransmitters	Eggs Whole grains such as oatmeal, brown rice, wholemeal bread Cruciferous vegetables (cabbage, broccoli, cauliflower, kale) Kidney beans, soya beans
Amino acids	Support production of neurotransmitters and aid memory and concentration	Organic meat Free-range poultry Fish Eggs Dairy products Nuts and seeds
Monounsaturated fats	Help keep blood vessels healthy, and support functions such as memory	Olive oil Peanuts, almonds, cashews, hazelnuts, pecans, pistachios Avocados
Antioxidants	Protect the brain cells from inflammation damage due to the presence of free radicals; improve cognitive functions and memory in older people	Dark chocolate (at least 70 per cent cocoa) Berries Pomegranates and juice Ground coffee Tea (especially green tea) Cruciferous vegetables Dark leafy greens Soya beans and products Nuts and seeds Nut butters, such as peanut butter and tahini
Water	Keeps brain hydrated to enable efficient chemical reactions	Tap water (especially "hard" water) Fruit and vegetables

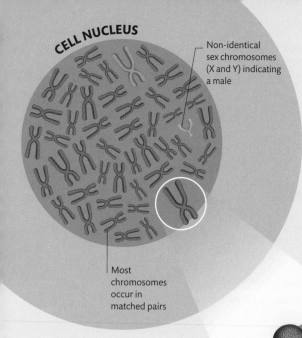

Non-identical sex chromosomes (X and Y) indicating a male

Most chromosomes occur in matched pairs

Chromosomes

We have around 20,000 genes, which are grouped into chromosomes. Each cell nucleus has 22 matched pairs of chromosomes (known as autosomes), plus a pair of sex chromosomes (identical XX chromosomes in females, or a non-identical pair, XY, in males).

ARE GENES ALWAYS ACTIVE?

Every DNA-bearing cell has a full set of genes, but many genes are normally only active in one part of the body, such as the brain, or at one stage of life, such as babyhood.

DNA and genes

The DNA molecule is a long, twisted strand formed from pairs of chemicals called bases – the "letters" of the genetic code – with a sugar-phosphate backbone at each edge. When cells divide, one half of the DNA goes into each new cell. In addition, we inherit one chromosome in each pair from our mother and one from our father, so each parent contributes half of our genes.

What is a gene?

Genes are sections of a long molecule called deoxyribonucleic acid (DNA), which contains the code that governs how our bodies develop and function. We inherit a mixture of genes from our parents. These genes produce proteins that shape physical traits, such as eye colour, or regulate processes such as chemical reactions. Their action turns these traits "on" or "off", or makes them more or less intense.

DNA helix is itself tightly coiled

Bases on one side of strand are paired with a complementary base on other side

Outer edge of each strand is made of sugar and phosphate molecules

Genetics and the brain

Genes govern the way our bodies, including the brain, develop and function. They work together with our environment to shape us throughout our life, from conception through to old age.

Four bases – adenine, thymine, guanine, and cytosine – are arranged in a particular sequence that encodes our genetic information

Adenine (red) always bonds with thymine (yellow)

MUTATION

When cells divide, the double-stranded DNA splits into single strands and each base is matched with a new complementary base, to form two new copies of the DNA. However, sometimes copying produces changes in the sequence. These may cause a gene to produce an altered protein or stop it working at all. Mutations may arise during life or may be inherited from parents.

Base pair

Backbone of DNA molecule

Mutation occurs when base pairs are changed during copying

New DNA strand made during cell copying

ERROR

AT LEAST ONE-THIRD OF ALL OUR **GENES** ARE ACTIVE PRIMARILY IN THE BRAIN

How faulty genes affect the brain

Genes do not directly control behaviour; instead, they govern the number and characteristics of nerve cells whose actions combine to produce our mental functions. For example, some genes influence the levels of neurotransmitters (see p.24), which in turn regulate functions such as memory, mood, behaviour, and cognitive skills. A faulty gene may fail to produce a protein needed for healthy brain function, or may increase the risk of a disorder such as Alzheimer's disease. Some faults can be inherited from parents; two inheritance patterns are shown here.

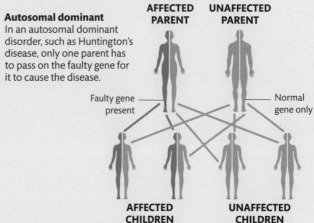

Autosomal dominant
In an autosomal dominant disorder, such as Huntington's disease, only one parent has to pass on the faulty gene for it to cause the disease.

AFFECTED PARENT

UNAFFECTED PARENT

Faulty gene present

Normal gene only

AFFECTED CHILDREN

UNAFFECTED CHILDREN

Autosomal recessive
In an autosomal recessive disorder, such as Tay-Sachs disease, the disorder only occurs if both parents pass on a faulty copy of the gene. Carriers have no disease themselves but can pass on the faulty gene.

CARRIER PARENT

CARRIER PARENT

Parent has one faulty and one healthy gene

Affected child has two copies of faulty gene

Unaffected child

Carrier children have one faulty and one healthy gene

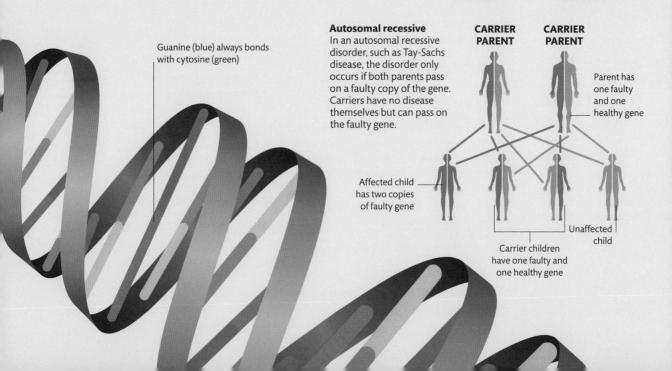

Guanine (blue) always bonds with cytosine (green)

WHEN IS THE SEX OF A FETUS FIXED?

Chromosomal sex is determined at the point of fertilization. Physical sexual differentiation occurs seven to 12 weeks after fertilization.

LARGER IN MALE BRAIN

Thalamus
This area, the "relay station" between the cortex and deeper brain structures, is larger in men than in women. The two sides of the thalamus are more likely to be connected in females, but the significance of this feature is not known.

LARGER IN FEMALE BRAIN

Corpus callosum
The corpus callosum, which links the brain's left and right hemispheres, has been found to be larger in females. It has been associated with greater cognitive skills in females, possibly because brain functions are shared between hemispheres, but not in males.

LARGER IN MALE BRAIN

Hippocampus
Males have a larger anterior (front) hippocampus, which governs acquiring and encoding new spatio-visual information, while females have a larger posterior hippocampus, which governs retrieval of existing spatio-visual knowledge.

Physical differences

Differences between males and females begin with the sex chromosomes at the moment of conception: XX for females and XY for males. In the uterus, the release of testosterone from the mother during gestation "masculinizes" a male fetus, triggering the growth of structural sex differences in both the brain and body. As we grow and develop, these differences will arise in many different brain structures (see right). Cognitive and skill differences between the sexes are present from childhood. Adult male brains are eight to 13 per cent larger, on average, than adult female brains. In addition, adult male brains also tend to vary more, in volume and cortical thickness, than female brains.

Male and female brains

Scientists have found that male and female brains show distinct physical differences. However, it is not always clear how these variations affect our attitudes, activities, and responses to our environment. Differences may arise from the way a brain is used in life as well as from its physical form.

ALL HUMAN EMBRYOS START LIFE WITH FEMALE BRAINS – EXTRA HORMONES ARE NEEDED TO CREATE A MALE

LARGER IN MALE BRAIN ♂

Hypothalamus
Certain areas governing male-typical sexual behaviour and responses to stress in the hypothalamus are larger in heterosexual males than in females or homosexual males.

LARGER IN MALE BRAIN ♂

Amygdala
The amygdala, involved in emotional responses, making decisions, and forming emotional memories, is slightly larger in males. However, differences in functions such as responses to negative versus positive emotional stimuli, are more significant.

Brain structures
There are several areas in which quantifiable physical differences have been identified between male and female adult brains. The main regions are shown here. How these differences can affect cognition and psychology are currently the matter of ongoing scientific research.

NON-BINARY BRAINS

Homosexual and transgender people have been found to have certain distinctive brain structures. For example, some parts of the hypothalamus (see above) differ in homosexual and heterosexual men, and the putamen (involved in learning and regulation of movement) has more grey matter in trans women than in cisgendered men.

NON-BINARY SYMBOL

Differences in function

Male and female brains differ in function as well as structure. Male brains seem to be more "lateralized" (with a greater difference in function between the left and right hemispheres). Males also vary more than females in cognitive ability. These variations are partly due to the structure of the "connectome" – the network of neural connections between parts of the brain (see below). They also result from the action of hormones, and external influences, throughout our life. In particular, our social environment and experiences continually shape our neural pathways, helping us perform male- or female-typical tasks.

Few connections cross hemispheres   Greater connectivity within hemispheres

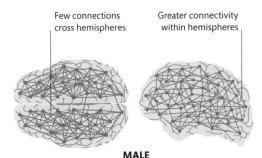

MALE

Many connections between hemispheres   Less connectivity within hemispheres

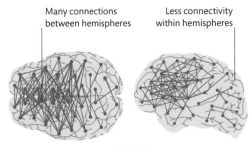

FEMALE

The connectome
One study, in which over 900 brains were imaged, found that male brains have greater connectivity within hemispheres, while female brains have denser connections between hemispheres. The males were found to be better at spatial processing, while the females scored higher on attention and memory for words and faces.

MUSICAL BRAINS

Playing music involves multiple parts of the brain. Studies comparing the brains of professional musicians and amateurs revealed that professional musicians had a greater volume of grey matter in brain areas related to motor, auditory, and visual-spatial reasoning. The study's findings show how the brain undergoes structural adaptations in response to the environment (dedicating hours to repetitive rehearsals with an instrument).

CHROMOSOMES

We inherit our chromosomes, which contain our DNA, from our parents (see pp.56–57). It's the chromosomes that, at the point of fertilization, determine the chromosomal sex of an embryo (XX for female and XY for male). Chromosomal abnormalities can also cause disease or developmental problems.

DNA

Some psychological traits, such as the tendency to develop depression, have been linked to particular genes – but they usually involve dozens or even hundreds of the genes acting together. The more of those genes a person inherits, the more likely they are to develop that trait.

THE **HIPPOCAMPUS** IN AN **ADULT BRAIN** MAKES AN ESTIMATED **700 NEW NEURONS** EVERY DAY

Genes versus environment

People are born with a DNA "template" inherited from their parents (see pp.56–57): this is the "nature" element influencing the brain's activities, such as cognitive ability and behaviour. Throughout a person's life, though, their networks of neurons (see pp.26–27) can adapt and change in response to physical and social experiences ("nurture"). Environmental influences, if strong and sustained, can alter brain structures and also influence the way that genes work – a process known as epigenetic change (see opposite).

WHEN DO EPIGENETIC CHANGES HAPPEN?

Epigenetic changes can be induced by environmental factors at any point in a person's life, from development in utero to old age.

Nature and nurture

The two fundamental influences on the brain, "nature" and "nurture", are sometimes seen as opposing forces. However, there is a dynamic interplay between them that goes on throughout a person's life.

NURTURE

PHYSICAL SURROUNDINGS

Studies on children have found that growing up poor or deprived can impair the development of areas related to memory, language processing, decision-making, and self-control. However, a safe, happy home, with interesting things to do, seems to reduce the harm.

STRESS LEVELS

Chronic emotional stress in children can impair development of the amygdala, hippocampus, and frontal lobes, leading to problems with memory, emotion, and learning. It restricts the action of genes regulating the growth of networks of neurons. However, moderate "positive" stress (fun) can aid learning.

DIET

A healthy diet (see pp.54–55) rich in omega-3 fatty acids, B vitamins, and antioxidants keeps blood vessels healthy, improving blood flow to the brain. These nutrients have also been linked to improving memory and maintaining cognitive functions in older people.

SOCIAL NETWORKS

Loneliness has been found to alter the production of neurotransmitters, so people perceive less reward from social contact and are more likely to misinterpret others' attitudes as threatening. However, maintaining close social ties can support memory and cognitive skills.

Epigenetic changes

Changes in the way genes are used (or expressed) that occur during a person's lifetime are called epigenetic changes. They affect gene function, rather than gene structure, and can be passed on to a person's children, although they may last for only a few generations. In the brain, they can influence functions such as learning, memory, reward-seeking, and response to stress. There are two main forms: methylation, in which a compound joins on to the DNA; and histone modification, which alters how tightly the DNA is coiled.

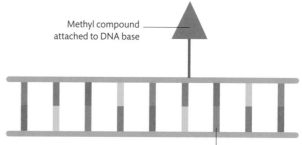

Methyl compound attached to DNA base

DNA methylation
In this process, a molecule of a methyl compound attaches to one of the bases in a gene's DNA sequence. The effect is to stop or restrict the activity of that gene.

Base pairs in most of sequence unchanged

STUDYING TWINS

Studies of twins reveal how much of a specific trait, such as intelligence quotient (IQ), is due to inheritance and how much is due to environment. Most twins grow up in the same home; however, identical twins share 100 per cent of their genes, while non-identical (fraternal) twins share only 50 per cent. If a trait is more evident in identical twins than in fraternal ones, or appears in identical twins who were separated at birth, it suggests that genetics has a stronger influence than environment.

BIOLOGICAL PARENTS

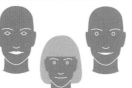

NON-ADOPTED TWIN

ADOPTIVE PARENTS

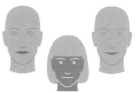

ADOPTED TWIN

BRAIN FUNCTIONS
AND THE SENSES

Sensing the world

To survive in our environment we must be able to react to, and interact with, stimuli produced by physical, chemical, and biological phenomena – sights, sounds, smells, tastes, and touches. Sensors in the body pick up these signals and send them to the brain for deciphering.

Senses

Each sense has its own set of detectors. Most are localized in a specific area of the body, except for touch, which is spread all over the skin, as well as inside the body. Although the neurons and receptors for each sense are largely dedicated to that sense alone, they can sometimes overlap. Sensory information continuously bombards the brain, but only a fraction of the input reaches consciousness. Even so, the "unnoticed" information can still guide our actions, particularly in the case of our sixth sense, proprioception, which relays information about the body's position in space.

YOUR SENSE OF SMELL IMPROVES WHEN YOU ARE HUNGRY

Touch
Thought to be the first sense to develop in the womb, touch neurons respond to pressure, temperature, vibration, pain, and light touch. Touch is how humans make physical contact with the environment and with each other.

Hearing
Sound waves in the air are collected by the ear and transmitted into the skull, where they are turned into electrical impulses by the cochlea. Hearing is the most developed of the senses at birth but is only complete by the end of the first year.

Sight
Sight involves sensors at the back of the eye that turn light into electrical signals. These are transported to the back of the brain, where they are converted into colours, fine details, and motion. We perceive objects in as little as half a second.

VISUAL CORTE[X]

SYNAESTHESIA

Synaesthesia is a condition where a stimulus may be interpreted by two or more senses at the same time. In its most common form, a person sees a number or word as a colour. Each synaesthete will have their own colour associations. Almost any combination of senses can be affected. Combinations of three or more senses are rare.

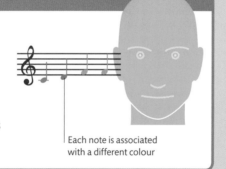

Each note is associated with a different colour

MOTOR
CORTEX

SOMATOSENSORY
CORTEX

PRIMARY TASTE
AREA

AUDITORY
CORTEX

SECONDARY TASTE
AREA

OLFACTORY
CORTEX

Proprioception
The brain is constantly processing information from the joints and muscles that tell it where the body is in space. It keeps us upright and allows us to make movements without conscious effort, such as walking up stairs.

Taste
Taste is important in determining what is safe and nutritious to eat. Taste receptors pick up only five basic tastes: sweet, salty, bitter, sour, and umami (savoury). We need our sense of smell to help identify a taste.

Smell
Despite having only 400 smell receptors, humans can detect up to a trillion different odours. Smell is important for survival as it warns us of hazardous substances or events, such as something burning. It also plays a key role in taste.

Sense areas of the cortex
Inputs from the sense receptors map to different areas of the brain's cortex. Although these areas are separate, they can often react to inputs from another sense. For example, visual neurons will respond better in low-light situations if they are accompanied by sound.

HOW MANY SENSES ARE THERE?

Including the six senses described here, scientists think there may be as many as 20 senses, based on the number of different receptor types in the body.

Seeing

The eye provides us with probably the most important of our five senses. It gathers the light reflected by an object and delivers this information to the brain via the optic nerve.

The structure of the eye

The eyeball is roughly 2.5 cm (1 in) in diameter. At the back of the eye is the retina, which contains light-sensitive cells that connect via neurons to the optic nerve. The space inside the eyeball is filled with a jellylike substance. The front of the eye contains a hole (the pupil), which has a clear lens behind it. Surrounding the pupil is a circle of coloured muscle, the iris, which controls how much light enters the eye. The cornea, a clear membrane, covers them and merges into the white outer membrane called the sclera.

WHY DO MY EYES CLOSE WHEN I SNEEZE?

When a nasal irritant triggers the brainstem control centre, it causes widespread muscle contractions, including those in the eyelids. This makes you blink momentarily.

Eyeball is encased by sclera

Crossed-over rays produce an inverted image on retina

Light rays start to refract (bend) as they pass from air and into cornea

LIGHT

Lens is like a bag of jelly that changes shape to help focusing

RETINA

CORNEA

PUPIL

IRIS

LENS

Iris is a ring of muscle

SCLERA

CHOROID

Choroid is a blood-rich layer that surrounds retina

Seeing things
The eye is capable of providing the brain with an enormous amount of detail about what it is looking at. However, the image the brain receives is inverted, so it has to be flipped before we can understand it.

Cornea is a transparent layer covering front of eye

1 Light enters the eye
Light passes through the cornea and into the eye through the pupil. The pupil is surrounded by a ring of coloured muscle, the iris, which can make the pupil contract or dilate to vary the amount of light entering.

2 Lens and focusing
Behind the iris is the lens, where the light rays are bent so the image forms on the retina. The lens is connected to muscles that allow it to change shape: it flattens for distant objects and thickens for close objects.

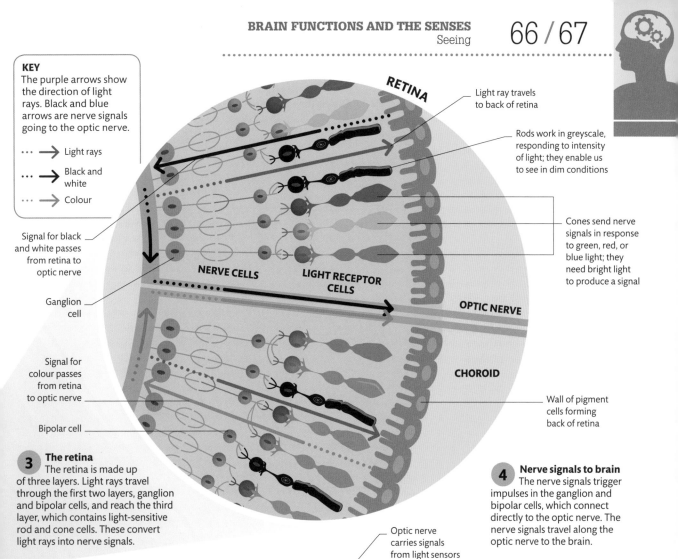

KEY
The purple arrows show the direction of light rays. Black and blue arrows are nerve signals going to the optic nerve.

···→ Light rays

···→ Black and white

···→ Colour

RETINA

Light ray travels to back of retina

Rods work in greyscale, responding to intensity of light; they enable us to see in dim conditions

Cones send nerve signals in response to green, red, or blue light; they need bright light to produce a signal

Signal for black and white passes from retina to optic nerve

Ganglion cell

NERVE CELLS

LIGHT RECEPTOR CELLS

OPTIC NERVE

CHOROID

Signal for colour passes from retina to optic nerve

Bipolar cell

Wall of pigment cells forming back of retina

3 The retina
The retina is made up of three layers. Light rays travel through the first two layers, ganglion and bipolar cells, and reach the third layer, which contains light-sensitive rod and cone cells. These convert light rays into nerve signals.

4 Nerve signals to brain
The nerve signals trigger impulses in the ganglion and bipolar cells, which connect directly to the optic nerve. The nerve signals travel along the optic nerve to the brain.

Optic nerve carries signals from light sensors to brain

OPTIC NERVE

YOUR EYEBALLS REMAIN THE SAME SIZE THROUGHOUT YOUR LIFE

THE BLIND SPOT

To connect to the brain, the nerve fibres of the retina must pass through the back of the eye to form the optic nerve. This creates a "blind spot" that has no photoreceptors. We do not notice this because each eye provides data about a scene and the brain uses information from the other eye to complete the picture.

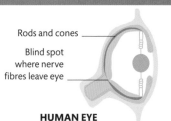

Rods and cones

Blind spot where nerve fibres leave eye

HUMAN EYE

The visual cortex

Nerve signals from the eye have to travel all the way through the brain before they reach the area dedicated to decoding this information. This area is called the visual cortex.

The structure of the cortex

The visual cortex occurs in both brain hemispheres and is further divided into eight main areas, each of which has a different function (see table, opposite). Signals travel from the retina (see pp.66–67) via the thalamus and lateral geniculate nucleus to the primary visual cortex (V1). The raw data then passes through various vision areas, contributing different details about shape, colour, depth, and motion before combining to form an image. Some areas provide information that helps with immediate recognition of familiar objects, others with spatial orientation or visual-motor skills.

3 Recognizing faces
Features that suggest a face are sent to the face-recognition area and amygdala, where they are searched for details that prompt recognition.

FRONTAL LOBE

Lateral geniculate nucleus forwards signals from retina to visual cortex

THALAMUS

Frontal lobe provides conscious recognition of faces

Amygdala processes facial expressions

VISUAL CORTEX

AMYGDALA

OPTIC NERVE

FACE RECOGNITION AREA

Rods and cones in retina convert light into nerve signals

Optic nerve carries nerve signals to brain

1 From eyeball to visual cortex
Data from the eyeball travels along the optic nerve until it reaches the optic chiasm (see below), where some of the data is sent to the opposite side of the brain. Signals then travel to the lateral geniculate nucleus, which forwards data to the visual cortex for processing.

KEY

• → Information from the eye

•→ Face recognition pathway

Stereoscopic vision

Our ability to see in 3D – known as stereoscopic vision – is produced by having both of our eyes looking straight ahead and moving together. As the eyes are slightly apart, different views are received from each, although they overlap to a small extent. The brain computes the spatial information from each eye to create an overall image, using previous experience to speed up the processing time and fill in any gaps.

Swapping sides

At a crossover point called the optic chiasm, nerve axons from the left side of each retina join and continue to the left visual cortex, and likewise with nerve axons from the right.

Lateral geniculate nucleus

LEFT HEMISPHERE

Half of signals travel to same hemisphere; other half cross over

View of object from left eye

LEFT VISUAL CORTEX

THALAMUS

RIGHT VISUAL CORTEX

RIGHT HEMISPHERE

Nerve axons split off after lateral geniculate nucleus and radiate to areas of visual cortex

Optic nerves converge at optic chiasm

View of object from right eye

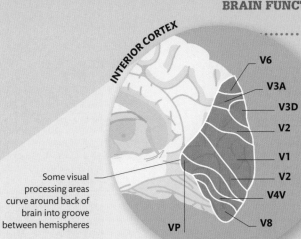

INTERIOR CORTEX

V6
V3A
V3D
V2
V1
V2
V4V
V8
VP

Some visual processing areas curve around back of brain into groove between hemispheres

THE **VISUAL CORTEX** IS VERY THIN – JUST **2 MM (0.08 IN)**

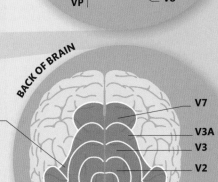

BACK OF BRAIN

Visual cortex, located in occipital lobe

V7
V3A
V3
V2
V4D
V1

2 The visual cortex
Nerve signals progress through the various layers of the cortex, each adding more information to the image. It takes half a second for the image to be assessed and become a conscious perception.

AREAS OF THE VISUAL CORTEX	
AREA	**FUNCTION**
V1	Responds to visual stimuli.
V2	Passes on information and responds to complex shapes.
V3A, V3D, VP	Registers angles and symmetry, and combines motion and direction.
V4D, V4V	Responds to colour, orientation, form, and movement.
V5	Responds to movement.
V6	Detects motion in periphery of visual field.
V7	Involved in perception of symmetry.
V8	Probably involved in processing of colour.

VISUAL FIELD OF LEFT EYE

Image formed by brain after it combines images from left and right eyes' visual fields

BINOCULAR VISUAL FIELD

VISUAL FIELD OF RIGHT EYE

FIELDS OF VISION

Animals such as primates have a large field of stereoscopic vision and can judge distances better than herbivores or most birds. However, they have a blind zone behind them that can only be seen by turning the head. Animals with eyes on the sides and top of the head have a wider field of 2D vision and greater all-around awareness.

RABBIT

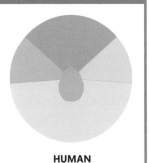

HUMAN

● Visual field of left eye   ● Visual field of right eye   ● Binocular visual field   ● Blind zone

How we see

Seeing is both a conscious and an unconscious action. Each type follows its own pathway in the brain. The conscious route helps recognize objects, while the unconscious route guides movement.

Cell area V1
Signals from the eyes are first received in the primary visual cortex (V1). Its neurons are sensitive to basic visual signals, including the orientation and direction of movement of objects and pattern recognition.

Cell area V2
In the secondary visual cortex (V2), some neurons improve on the images from V1, sharpening the lines and edges of complex shapes. Other neurons refine the initial interpretation of the colour of objects.

Cell area V3
Visual area 3 (V3) is involved in analysing angles, position, depth, and the orientation of shapes. It also helps to process the direction and speed of objects. A few cells are also sensitive to colour.

VISUAL CORTEX PATHWAY

Following the path

As visual information is processed by the layers of the visual cortex (see pp.68–69), it splits into two pathways known as the upper, or dorsal, route and the lower, or ventral, route. There is some uncertainty about where the split occurs, but the dorsal route handles our spatial awareness of where we are and how we move in relation to things around us, while the ventral route helps us identify, categorize, and recognize what we see. The dorsal route is important in assessing important situations, particularly if instant action is required to avoid danger, such as moving away from a flying object. When this happens, the ventral route is relegated to a secondary position since the information it carries is not critical.

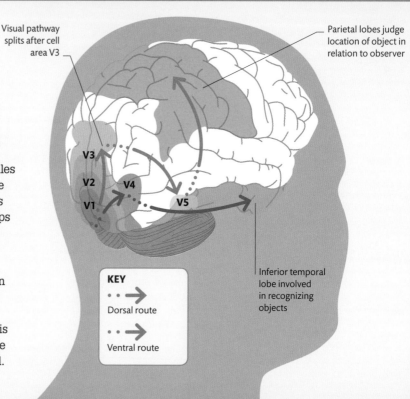

Visual pathway splits after cell area V3

Parietal lobes judge location of object in relation to observer

V3
V2
V4
V1
V5

Inferior temporal lobe involved in recognizing objects

KEY

Dorsal route

Ventral route

Cell area V5
The middle temporal area (V5) judges the overall direction of motion of an object rather than that of its component parts. For example, it processes the general direction of a flock of birds, rather than the movement of a single bird. It also analyses the motion of our own body.

Parietal lobe
The parietal lobe gauges the depth and position of an object in relation to the observer. This allows the person to take immediate action, such as ducking an object coming towards them at speed.

"WHERE" PATHWAY (DORSAL ROUTE)

Unconscious vision
The dorsal route carries visual information to the parietal lobes, passing through areas that calculate an object's location, timing, and motion and make a plan in relation to it. All this happens without any conscious thought.

Conscious vision
The ventral route adds more information to the object, such as colour and shape. The information goes to the temporal lobe, where it is matched to visual memories to aid recognition. This is where the visual stimulus becomes a conscious perception.

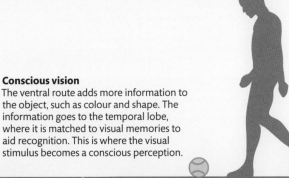

"WHAT" PATHWAY (VENTRAL ROUTE)

Cell area V4
Visual area 4 (V4) is involved in the perception of colour, texture, orientation, form, and movement. This region contains the majority of colour-sensing neurons and is important in interpreting the space between objects.

Inferior temporal lobe
Signals are forwarded to the fusiform gyrus of the inferior temporal lobe, which is involved in recognizing complex shapes, objects, and faces. In conjunction with the hippocampus, it helps with the formation of new memories.

WHAT IS PROSOPAGNOSIA?
This is the inability to recognize faces, even of close family, usually due to damage to the inferior temporal lobe. Those affected have to learn to recognize people in other ways.

Perception

Given that visual processing happens in microseconds, it is not surprising that our brain sometimes struggles to make sense of the information sent by our eyes, and so makes us doubt what we are seeing.

Brain is so drawn to faces that even pictures are studied

Processing a scene

When we look at a scene, we are not really taking it all in. Instead, the eyes repeatedly scan a sequence of thumbnail-sized areas that the brain considers points of interest. The rest of the scene blurs until attention falls onto a new area. Faces tend to be the main focus of a scene – the brain is programmed to look for faces, hence the tendency to see them in the unlikeliest of places, such as the scorch marks on a slice of toast. While details of the target objects are being scrutinized, the conscious brain puts together the story of the scene, complete with the context of each object.

Openings are scanned, perhaps for possibility of intruders

Pointing draws attention to an object and makes it worthy of a look

Eye passes straight across floor, pausing briefly at a potential obstacle, but not stopping long enough to see it

Scanning for details

Looking at a complex picture, such as this café scene, activates processes that distinguish target objects, such as people, from the background and then selects which bits of the target to focus on.

Brain looks for clues about relationships by looking at individual faces and interplay between characters

WHY DO WE SEE FACES IN INANIMATE OBJECTS?

Pareidolia (seeing faces where there are none) may be a survival instinct that ensures we are vigilant for the unfriendly features of an enemy or predator.

Illusions

An illusion occurs when what the eye sees is interpreted by the brain in a way that does not match up with the physical reality of the actual image. With so many competing signals going to the brain, it tends to look for familiar patterns. It also tries to predict what will happen next to compensate for the slight time delay between stimulus and perception. Both these facts can lead to our brain misinterpreting visual stimuli. Illusions fall into three main classes: physiological, cognitive, and physical.

HERMANN GRID

Physiological

Physiological illusions are thought to arise from excessive or competing stimuli, such as brightness, colour, movement, and position. In this grid, grey spots seem to appear at the intersections as your eyes flick over them, but vanish when you stare at them.

KANIZSA'S TRIANGLE

Cognitive

Cognitive illusions happen when the brain makes assumptions about movement or perspective when viewing an object. Sometimes these can lead to the brain switching between two different images or seeing a shape that is not there.

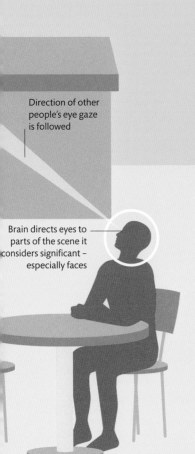

Direction of other people's eye gaze is followed

Brain directs eyes to parts of the scene it considers significant – especially faces

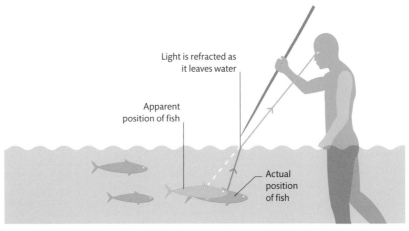

Light is refracted as it leaves water

Apparent position of fish

Actual position of fish

REFRACTION

Physical

Physical illusions are caused by the optical properties of the physical environment, particularly water. The brain cannot take account of the way that light bends as it passes between water and air, so it sees the fish as further back than it actually is.

SOME **MAMMALS AND BIRDS** ARE ALSO **FOOLED** BY OPTICAL **ILLUSIONS**

How we hear

The world is full of noise. It travels as sound waves through the air until it reaches our ears. There, they are turned into electrical impulses and sent to the brain for decoding into meaningful sounds.

Picking up sound

Hearing involves the conversion of a sound wave into an electrical impulse that the brain can interpret. Sound waves are carried from the outer to the middle ear, where they cause a series of bones and membranes to vibrate. These vibrations then reach the cochlea, where they become electrical impulses. These are passed to the brainstem and thalamus, where direction, frequency, and intensity are perceived. The data is then sent for processing by the left and right sides of the auditory cortex. The left side identifies the sound and gives it meaning, while the right side assesses the quality of the sound.

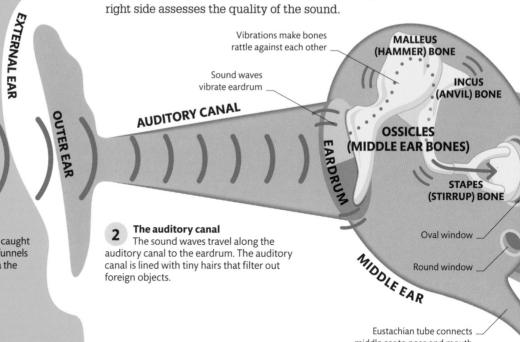

EXTERNAL EAR

OUTER EAR

AUDITORY CANAL

EARDRUM

MIDDLE EAR

Sound waves travel through air

Vibrations make bones rattle against each other

Sound waves vibrate eardrum

MALLEUS (HAMMER) BONE

INCUS (ANVIL) BONE

OSSICLES (MIDDLE EAR BONES)

STAPES (STIRRUP) BONE

Oval window

Round window

Eustachian tube connects middle ear to nose and mouth

1 **The outer ear**
Sound waves are caught by the outer ear, which funnels them inside the head via the auditory canal.

2 **The auditory canal**
The sound waves travel along the auditory canal to the eardrum. The auditory canal is lined with tiny hairs that filter out foreign objects.

3 **The eardrum**
The eardrum, or tympanic membrane, is a thin layer of fibrous tissue that forms a barrier between the outer ear and the middle ear. It vibrates when the sound waves travelling up the auditory canal hit it.

4 **Ossicles**
Vibrations are passed through the eardrum to a set of connected bones called ossicles – the malleus, incus, and stapes bones. The stapes bone pushes and pulls on another membrane, called the oval window. This transmits sound to the inner ear.

FILTERING OUT NOISE

On a busy street there are lots of conflicting sounds, yet you can still hear someone talking next to you. This is because the primary auditory cortex can filter out unnecessary sounds and boost the signals it wants to hear. It does this by dampening the response to sustained sounds, such as traffic, while enhancing more dynamic sounds, such as speech, and actively listening to them.

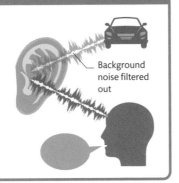

Background noise filtered out

9 **The primary auditory cortex**
After intermediate processing in the thalamus, the characteristics of each sound are interpreted by the primary auditory cortex, which works with other cortical areas to identify the type of sound.

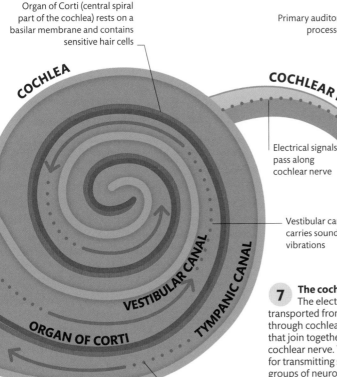

Organ of Corti (central spiral part of the cochlea) rests on a basilar membrane and contains sensitive hair cells

Primary auditory cortex processes sound

THALAMUS

COCHLEA

COCHLEAR NERVE

Electrical signals pass along cochlear nerve

Vestibular canal carries sound vibrations

VESTIBULAR CANAL

TYMPANIC CANAL

ORGAN OF CORTI

BRAINSTEM

Specialized cells at top of brainstem help determine direction of sounds

7 **The cochlear nerve**
The electrical signals are transported from each hair cell through cochlear nerve endings that join together to form the cochlear nerve. This is responsible for transmitting signals to specialized groups of neurons in the brainstem.

8 **The thalamus**
Signals are first received in the brainstem. From here, they travel up to specialized neurons in the thalamus for processing. These signals are then sent to the primary auditory cortex, which also feeds information back to the thalamus.

Vibrations return to round window

INNER EAR

5 **The cochlea**
The cochlea contains three fluid-filled ducts. Vibrations travel along the vestibular canal as wavelike movements that are transferred to the basilar membrane of the organ of Corti. Residual vibrations return along the tympanic canal to the round window.

6 **The organ of Corti**
The movement of the basilar membrane bends sensitive hair cells in the organ of Corti (see p.76), which is the main organ of hearing. The hair cells convert this movement into electrical signals.

THE STAPES IS THE SMALLEST BONE IN THE BODY

Perceiving sound

Every sound is made up of a number of different components. The brain has to take all the details of its frequencies, intensity, and rhythm to process, identify, and remember the sound.

This area receives signals from low-frequency sounds

Corresponds to apex of cochlea

PRIMARY

SECONDARY

TERTIARY

Corresponds to base of cochlea

Receives signals from high-frequency sounds

The auditory cortex
The auditory cortex is the main processing centre for sound. It is located in the temporal lobe, just below the temples on either side of the head.

Primary auditory cortex identifies frequency and intensity of sounds

Secondary auditory cortex interprets complicated sounds, such as language

AUDITORY CORTEX

Hair cells are disturbed when basilar membrane vibrates

More flexible part of basilar membrane vibrates more easily

Base of cochlea transmits low-frequency sounds

Organ of Corti is main organ of hearing

Tertiary auditory cortex integrates hearing with other sensory systems

Apex of cochlea transmits high-frequency sounds

500 Hz

1,000 Hz

2,000 Hz

4,000 Hz

8,000 Hz

16,000 Hz

BASILAR MEMBRANE

COCHLEA

Row of hair cells

Inside the auditory cortex

Signals from the thalamus (see p.75) are sent to the primary auditory cortex, which is divided into sections that respond to a range of frequencies. Some of these sections focus on intensity rather than frequency, while others pick up more complex and distinctive sounds, such as whistles, bangs, or animal noises. Signals then pass to the secondary auditory cortex, which is thought to focus on harmony, rhythm, and melody. The tertiary auditory cortex integrates all the signals to give an overall impression of the sounds picked up by the ears.

The cochlea
Areas along the curl of the cochlea respond to different frequencies of sound, from high-pitched at the apex to low bass notes at the base. These are mirrored by corresponding areas in the auditory cortex.

Music and the brain

Music engages many areas of the brain. As well as processing the sounds, listening to music also triggers the memory and emotion centres in the brain, while recalling lyrics involves the language centres. Performing music makes even greater demands; the visual cortex is stimulated by reading music, the frontal lobe is involved in planning actions, and the motor cortex coordinates movement. Musicians are known to have a greater ability to use both hands because music requires co-ordination of motor control, somatosensory touch, and auditory information. Unlike listeners, who process music in the right hemisphere, professional musicians use the left. They also have a thicker corpus callosum (the region linking the two hemispheres) and tend to have larger auditory and motor cortices.

30,000
THE NUMBER OF **FIBRES** THAT MAKE UP THE **AUDITORY NERVE**

Mapping music

Scans show that several areas of the brain are active when listening to music, and even more are involved when you are playing an instrument or dancing.

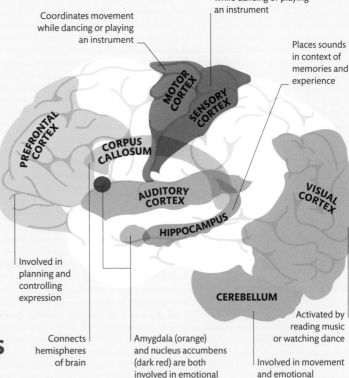

Coordinates movement while dancing or playing an instrument

Processes touch sensations while dancing or playing an instrument

Places sounds in context of memories and experience

MOTOR CORTEX

SENSORY CORTEX

PREFRONTAL CORTEX

CORPUS CALLOSUM

AUDITORY CORTEX

VISUAL CORTEX

HIPPOCAMPUS

CEREBELLUM

Involved in planning and controlling expression

Connects hemispheres of brain

Amygdala (orange) and nucleus accumbens (dark red) are both involved in emotional reactions to music

Activated by reading music or watching dance

Involved in movement and emotional reaction to music

HIGHS AND LOWS

Humans can detect a good range of frequencies, but other animals can hear things far beyond our limits. Animals such as bats and dolphins use high frequencies in echolocation, while elephants and whales produce low rumbles that can travel long distances. Humans are most sensitive to frequencies between 2 kHz and 5 kHz, which do not require great intensity to be heard. Young people have the best hearing range, from 20 Hz to 20 kHz, but there is a gradual loss of higher frequencies with age, with older people having a limit of around 15 kHz.

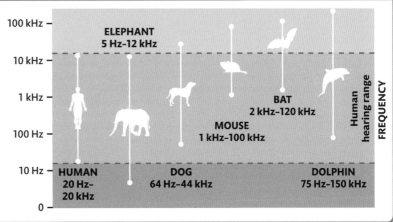

100 kHz

10 kHz

1 kHz

100 Hz

10 Hz

0

ELEPHANT 5 Hz–12 kHz

BAT 2 kHz–120 kHz

MOUSE 1 kHz–100 kHz

Human hearing range

FREQUENCY

HUMAN 20 Hz– 20 kHz

DOG 64 Hz–44 kHz

DOLPHIN 75 Hz–150 kHz

OLFACTORY EPITHELIUM

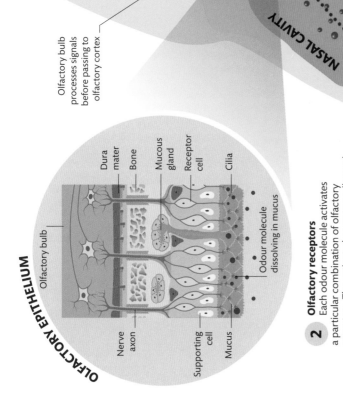

Olfactory bulb

Dura mater

Bone

Mucous gland

Receptor cell

Cilia

Nerve axon

Supporting cell

Mucus

Odour molecule dissolving in mucus

2 Olfactory receptors

Each odour molecule activates a particular combination of olfactory receptors. The activated receptor cells send impulses up through nerve axons to the olfactory bulb for processing.

1 Smell enters the nose

Odour molecules are drawn up through the nose and warmed to enhance the scent. The molecules dissolve in mucus produced by the olfactory epithelium and stimulate cilia that are connected to receptor cells.

Airborne odour molecules enter nostril

12 MILLION

THE NUMBER OF **OLFACTORY CELLS** IN THE **HUMAN BODY**

---

3 Inside the brain

Signals then travel along the olfactory tract to the olfactory cortex. The cortex is located in the limbic system, which is responsible for emotions and memory. Signals are also sent to the amygdala and orbitofrontal cortex.

OLFACTORY CORTEX

Olfactory cortex further processes signals sent by olfactory bulb

Amygdala sends warning messages if odour is associated with danger

AMYGDALA

Olfactory tract, a bundle of nerves that carries signals from olfactory bulb to olfactory cortex

ORBITOFRONTAL CORTEX

Orbitofrontal cortex involved in decision-making and emotions, as well as processing smells

Olfactory bulb processes signals before passing to olfactory cortex

OLFACTORY BULB

Receptor cell nerve axons detect odour and send information to olfactory bulb

NASAL CAVITY

Capturing a scent

When we inhale, odour molecules drift into the nose and activate receptor cells in the nasal cavity, triggering a reflex to breathe in more deeply. In the nasal cavity, the odours dissolve in the mucus that covers a sheet of neurons and supporting cells called olfactory epithelium. The molecules spread through the mucus to hairlike structures called cilia that are attached to receptor cells. These cells send signals to the olfactory bulb – a structure located in the forebrain that forms part of the brain's limbic system. Data is then sent to various parts of the brain, particularly the olfactory cortex.

Smell

Identifying a smell out of the many odours in the world around us involves the olfactory system, which isolates different chemicals and then passes signals on to the brain to determine whether they are "good" or "bad".

What makes a smell?

How we identify smells is still a matter of debate. Research suggests that most odours fall into ten groups – or primary odours – each of which alerts us to something in the environment. Most smells are made up of a combination of these groups. Smell is a key part of survival, telling us whether something is safe or dangerous.

Fragrant
Light, natural scents such as flowers, grasses, and herbs, typically used in perfumery.

Fruity
Typically includes warm, ripe fruits and other fresh scents that have a sense of smoothness on the nose.

Citrus
Separate from other fruits, citrus has fresh, clean, acidic aromas with a touch of sweetness.

Woody and resinous
Earthy, natural smells, such as compost, fungi, spices, cedar, pine, and mould.

Chemical
Includes synthetic, medicinal, solvent, and gasoline odours that are easily identifiable.

Sweet
Warm, rich, sugary smells with a touch of creaminess, including chocolate, malt, and vanilla.

Minty
Cool, fresh, and invigorating, epitomised by mint, eucalyptus, and camphor.

Toasted and nutty
Slightly burned and caramelized with warm and fatty overtones, such as popcorn and peanut butter.

Pungent
Often unpleasant smells such as manure or sour milk, also onions, garlic, and pickles.

Decayed
Beyond pungent are the odours of rotting food, sewage, household gas, and other "sickening" substances.

WHY DO SMELLS TRIGGER MEMORIES?

Unlike our other senses, smells bypass the thalamus and go straight to the limbic system. Emotions and memories are processed and stored here, especially in the amygdala.

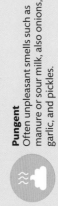

SMELLY OR SWEET?

Dimethyl sulphide (DMS) is a very smelly compound. A whiff of the raw chemical can make you wonder whether something is rotting or if a pungent cheese is in the room. However, flavour chemists find it useful in creating all sorts of tastes. It is used in meat, seafood, milk, egg, wine, beer, vegetable, and fruit flavourings, usually at minuscule concentrations.

Taste

Fuelling the body requires the intake of nourishing foods and liquids. Choosing what is safe to eat is largely influenced by our senses of taste and smell.

Picking up taste

Taste is actually a limited sense; there are only five basic tastes that can be detected (see right). Like smell, taste is a chemosense. Chemical substances in food are picked up by the taste buds, which are mainly found on the tongue. Receptor cells, housed in structures called microvilli within the taste bud, detect these chemicals and send signals to the brain for processing.

The five basic tastes

Taste is an evolutionary adaptation for survival. Being able to determine whether something is nutritious or potentially poisonous before taking it into the body is enormously important. So far, only five basic tastes have been discovered, although there may be others.

Sweet
Signals the presence of carbohydrates, which are sources of vital sugars.

Salty
Detects chemical salts and minerals that are needed by the body.

Sour
Warns against foods that may be unripe or going off.

Bitter
Poisons and other toxins are often bitter or unpalatable.

Umami
Detects glutamate salts and amino acids, which are found in meat, cheese, and other aged or fermented foods.

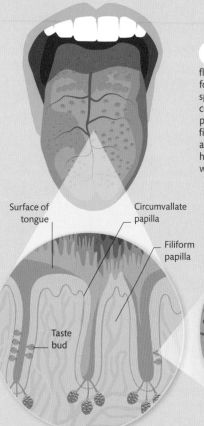

1 **Tongue**
The tongue is a strong, flexible muscle. It is used to push food around the mouth and for speech. Its upper surface is covered in tiny projections called papillae. Most of the papillae are filiform, or threadlike, structures and contain no taste buds. They help grip and wear down food while it is being chewed.

Surface of tongue

Circumvallate papilla

Filiform papilla

Taste bud

Taste pore

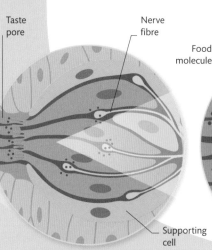

Nerve fibre

Food molecule

Supporting cell

Microvilli contain receptor proteins, which bind with chemicals in food

Neuron

Gustatory receptor cell

2 **Papillae**
As well as filiform papillae, the tongue has fungiform (mushroomlike), foliate (leaflike), and circumvallate (wall-like) papillae, which all contain taste buds. Most taste buds are found in the foliate papillae on the back and sides of the tongue.

3 **Taste buds**
A taste bud is a collection of 50–100 cells that are clustered together like segments in an orange. They are located in the walls of papillae. One end of each cell protrudes out of the bud, where it gets washed with saliva containing food molecules.

4 **Taste bud cells**
When food molecules hit the cells, they interact with either receptor proteins or porelike proteins called ion channels. This causes electrical changes in the cell, which prompt neurons at the base of the cell to send signals to the brain.

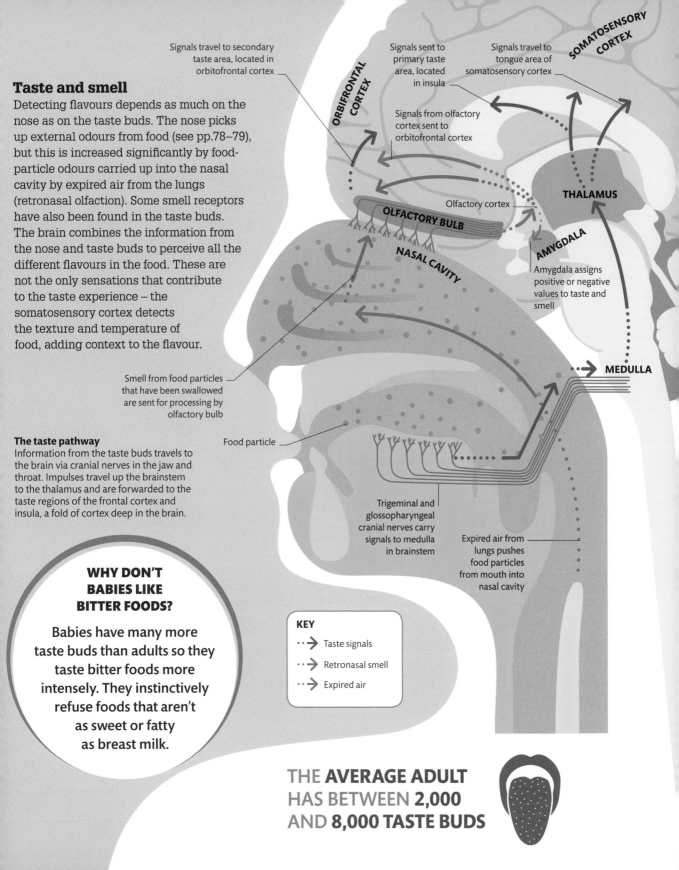

Taste and smell

Detecting flavours depends as much on the nose as on the taste buds. The nose picks up external odours from food (see pp.78–79), but this is increased significantly by food-particle odours carried up into the nasal cavity by expired air from the lungs (retronasal olfaction). Some smell receptors have also been found in the taste buds. The brain combines the information from the nose and taste buds to perceive all the different flavours in the food. These are not the only sensations that contribute to the taste experience – the somatosensory cortex detects the texture and temperature of food, adding context to the flavour.

The taste pathway

Information from the taste buds travels to the brain via cranial nerves in the jaw and throat. Impulses travel up the brainstem to the thalamus and are forwarded to the taste regions of the frontal cortex and insula, a fold of cortex deep in the brain.

Signals travel to secondary taste area, located in orbitofrontal cortex

Signals sent to primary taste area, located in insula

Signals travel to tongue area of somatosensory cortex

Signals from olfactory cortex sent to orbitofrontal cortex

ORBIFRONTAL CORTEX

SOMATOSENSORY CORTEX

Olfactory cortex

THALAMUS

OLFACTORY BULB

AMYGDALA

Amygdala assigns positive or negative values to taste and smell

NASAL CAVITY

Smell from food particles that have been swallowed are sent for processing by olfactory bulb

Food particle

Trigeminal and glossopharyngeal cranial nerves carry signals to medulla in brainstem

MEDULLA

Expired air from lungs pushes food particles from mouth into nasal cavity

WHY DON'T BABIES LIKE BITTER FOODS?

Babies have many more taste buds than adults so they taste bitter foods more intensely. They instinctively refuse foods that aren't as sweet or fatty as breast milk.

KEY

- ⋯→ Taste signals
- ⋯→ Retronasal smell
- ⋯→ Expired air

THE **AVERAGE ADULT** HAS BETWEEN **2,000** AND **8,000 TASTE BUDS**

LIGHT BREEZE

TEMPERATURE CHANGE

BRUSH OF A FEATHER

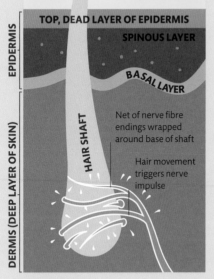

EPIDERMIS

TOP, DEAD LAYER OF EPIDERMIS

SPINOUS LAYER

BASAL LAYER

DERMIS (DEEP LAYER OF SKIN)

HAIR SHAFT

Net of nerve fibre endings wrapped around base of shaft

Hair movement triggers nerve impulse

Free nerve endings extend into skin's surface layer

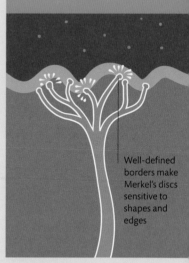

Well-defined borders make Merkel's discs sensitive to shapes and edges

Root hair plexus
Nerves wrapped around the base of a hair shaft are triggered by things that have not touched the skin, such as air currents or objects that brush against the hair.

Free nerve endings
Extending up into the spinous layer of the epidermis, these bare, rootlike nerve endings are sensitive to cold, heat, light touch, and pain.

Merkel's discs
Found slightly lower than free nerve endings, Merkel's discs are particularly dense in the lips and fingertips. They respond to light touch.

Touch

The skin is the biggest organ of the body and also the largest sense organ. Packed with sensors, it enables us to experience a wide variety of sensations, as well as an awareness of where we are.

Receptors in the skin

Skin sensors consist of receptors connected by axons. Found at various levels in the skin, there are around 20 types that respond to different sorts of stimuli. The receptors register mechanical, thermal, and, in some cases, chemical stimuli and convert them into electrical signals. These travel up peripheral nerves to the spinal cord, then to the brainstem, and finally to the somatosensory cortex, where they are translated into a touch.

TYPES OF RECEPTOR	FUNCTION
Mechanoreceptors	Sensory receptors that respond to mechanical pressure or distortion. This can range from a light touch to deep pressure.
Proprioceptors	Receptors that receive stimuli from within the body, particularly in relation to position and movement.
Nociceptors	Sensory neurons that respond to damaging stimuli by sending "possible threat" signals to the spinal cord and the brain.
Thermoreceptors	Specialized nerve cells that are able to detect differences in temperature. They are found all over the skin and in some internal areas.
Chemoreceptors	Extensions of the peripheral nervous system that respond to changes in blood concentrations to maintain homeostasis (see pp.90–91).

GENTLE TOUCH

Fluid-filled receptors extend into upper dermis

Meissner corpuscles
These receptors are rapidly adapting, meaning that they respond quickly to stimulation but stop firing if the stimulus continues. This gives precise information.

FIRM MASSAGE

Enlarged, encapsulated receptor

Ruffini endings
Also known as bulbous corpuscles, these soft, capsulelike cells – located deep in the dermis – respond if the skin or joints are stretched or distorted by pressure.

VIBRATION

Large, covered receptor at base of dermis

Pacinian corpuscle
The deepest and largest type of touch receptor, these rapidly acting mechanoreceptors respond to sustained pressure as well as vibration.

The somatosensory cortex

All information from touch receptors is processed in the somatosensory cortex. This area sits across the top of the brain like a hairband. Data from the right side of the body travels to the left side of the brain, and vice versa. Each part of the body maps to its own area of the cortex.

Touch map
Areas of the body rich in touch receptors, such as the hands, require more processing than others, so they take up a greater proportion of the somatosensory cortex.

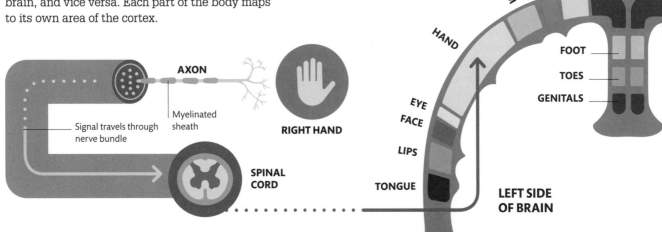

AXON

Myelinated sheath

Signal travels through nerve bundle

RIGHT HAND

SPINAL CORD

ARM · HEAD · TRUNK · LEG

HAND

EYE
FACE

LIPS

TONGUE

FOOT

TOES

GENITALS

LEFT SIDE OF BRAIN

Proprioception

The body has its own sense of where it is and how it is moving in space. This process happens almost unconsciously, making it, in essence, the body's sixth sense.

Body position sense

Inside muscles, tendons, and joints are movement receptors called proprioceptors. Every time we move, these receptors measure changes in length, tension, and pressure that relate to that movement and send impulses to the brain. The information is processed and a decision is made to stop moving or change position. Messages are then relayed back to the muscles to carry out the decision. All this happens without us having to think about it.

Types of proprioception

Most of the information our brain receives about body position is processed unconsciously, such as how we are constantly adjusting the position of our body to maintain balance. However, proprioceptive information can become conscious if it requires us to make a decision – for example, refining muscle movement to make a voluntary, skilled movement.

Proprioception pathways

Conscious proprioception signals travel up the brainstem to the thalamus and end at the parietal lobe, which is part of the cerebral cortex. The unconscious pathway loops back to the cerebellum, which controls movement.

Knowing your place

Physical self-awareness comes from a combination of proprioception with other sensations: a sense of force, a sense of effort or weight, sight, and information from the balance organs in the ears.

PERIPHERAL NERVE

Nerve signal from proprioceptors

Stretch receptors in skin, muscles, and joints send information about position of body parts

SPINAL COLUMN

Signals travel along spinal column to brain

Parietal lobe

Inner ear sends information about rotation, acceleration, and gravity

Eyes send visual information about position

Input from pressure and tension sensors in arms

Parietal lobe

Thalamus

Cerebellum

Unconscious pathway

Conscious pathway

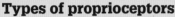

Types of proprioceptors

The body contains a variety of proprioceptors, and the combined information from these receptors helps the brain to construct an overall picture of the body's position. There are three main types of proprioceptors: muscle spindle fibres, which are embedded in our muscles; Golgi tendon organs, which are located at the junction between tendons and muscles; and joint receptors, which attach to our joints. Special receptors in the skin can also detect stretch (see p.83).

GROWTH SPURTS CAN CONFUSE THE BRAIN, AS IT CANNOT KEEP UP WITH CHANGES IN LIMB DIMENSIONS

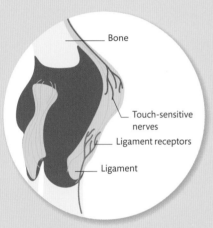

Bone

Touch-sensitive nerves

Ligament receptors

Ligament

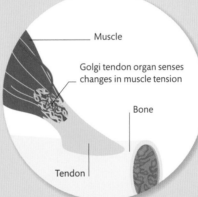

Muscle

Golgi tendon organ senses changes in muscle tension

Bone

Tendon

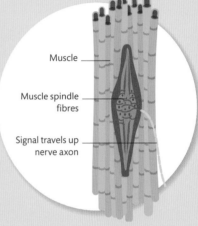

Muscle

Muscle spindle fibres

Signal travels up nerve axon

Joint receptors
Nerve endings within our joints detect the joints' position. The receptors help prevent damage through over-extension as well as detecting position in normal motion.

Tendon receptors
Golgi tendon organs are found within the tendons at the ends of muscles. They monitor muscle tension to ensure we do not overstretch the muscles.

Muscle receptors
Muscles have position sensors called spindle fibres within them. As they stretch, the spindles send information to the brain about the positions of the muscles.

THE PINOCCHIO ILLUSION

Sometimes proprioception can be confused, making the body feel like something is happening when it is not. One such effect is called the Pinocchio illusion. A vibrator is fixed to a person's bicep. If the person holds her nose while the vibrator is turned on, she will feel as though her arm is moving away from her nose. It happens because the vibrator stimulates the muscle spindle fibres in the biceps in the same way as if the muscle was stretching. Because the fingers are still touching the nose, it feels as if the nose is growing out from the face.

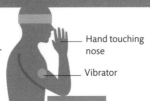

Hand touching nose

Vibrator

Before stimulation
At rest, the brain is aware that the fingers are touching the nose but there is no movement of the arm.

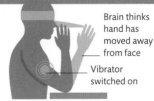

Brain thinks hand has moved away from face

Vibrator switched on

During stimulation
Vibrations tell the brain that the arm is moving, creating a sensation that the nose is growing outwards.

Feeling pain

Although unpleasant, pain is a useful warning sign that something isn't right with the body and that we need to act quickly to avoid further injury.

WHO FEELS THE MOST PAIN?

Women feel pain more intensely than men because they have more nerve receptors in their bodies.

Pain signals

Pain receptors are located all over the body, and respond to heat, cold, overstretching, vibration, and chemicals released by wounds. Electrical signals are sent from the site of injury to the spinal cord, where they cross over and travel to the opposite side of the brain to the injury. If sudden, strong pain is experienced, a reflex reaction occurs (see p.101) within the spinal cord to make the limb pull away from whatever is causing the pain, even before we are aware of it.

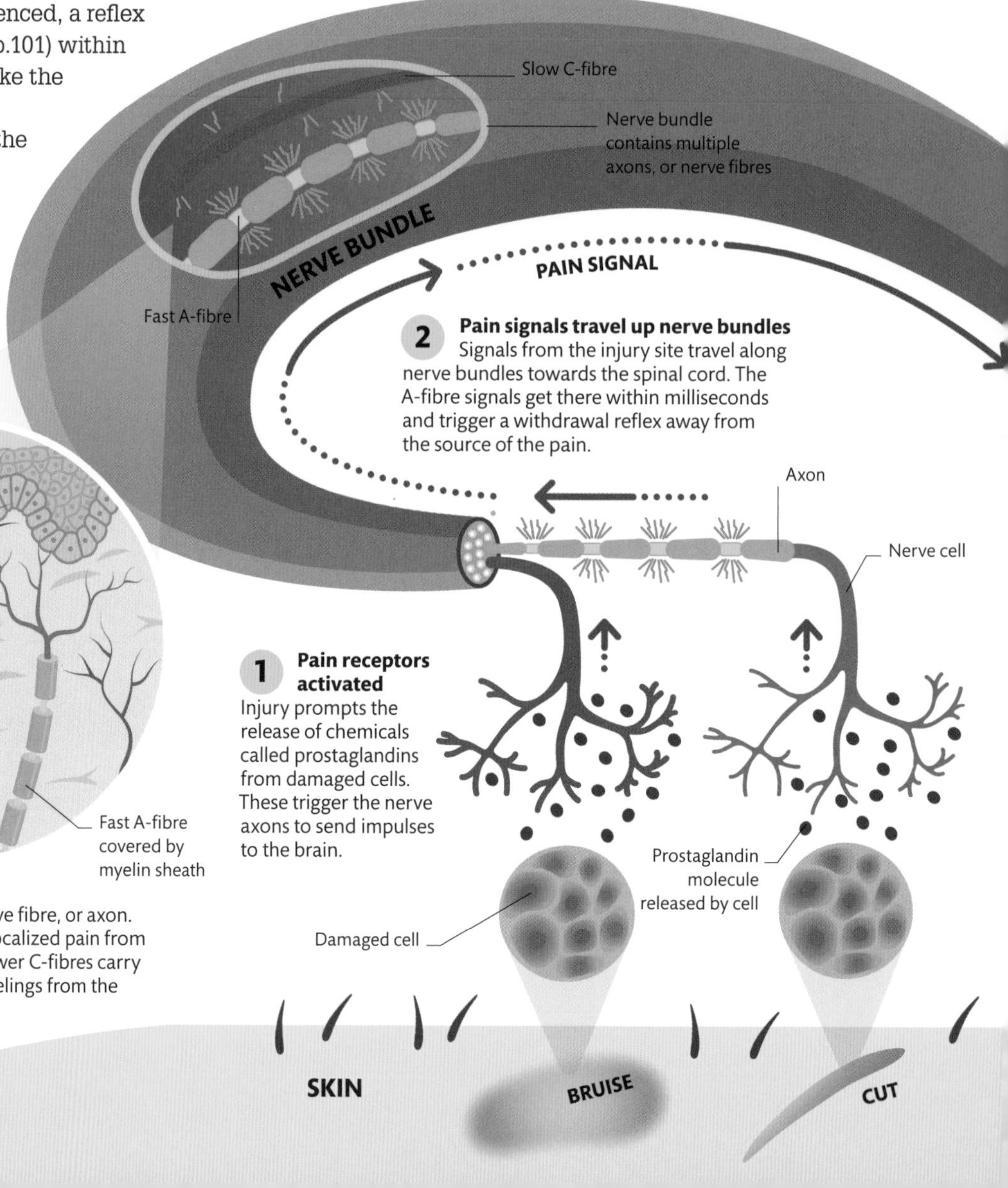
Slow C-fibre
Nerve bundle contains multiple axons, or nerve fibres
NERVE BUNDLE
Fast A-fibre
PAIN SIGNAL
Axon
Nerve cell
Prostaglandin molecule released by cell
Damaged cell
SKIN
BRUISE
CUT

2 Pain signals travel up nerve bundles
Signals from the injury site travel along nerve bundles towards the spinal cord. The A-fibre signals get there within milliseconds and trigger a withdrawal reflex away from the source of the pain.

1 Pain receptors activated
Injury prompts the release of chemicals called prostaglandins from damaged cells. These trigger the nerve axons to send impulses to the brain.

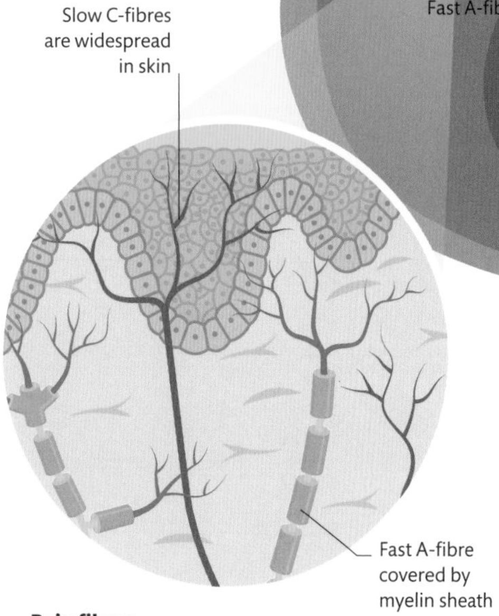
Slow C-fibres are widespread in skin
Fast A-fibre covered by myelin sheath

Pain fibres
There are two types of nerve fibre, or axon. Fast A-fibres carry sharp, localized pain from an injury such as a cut. Slower C-fibres carry the more persistent dull feelings from the area around the injury.

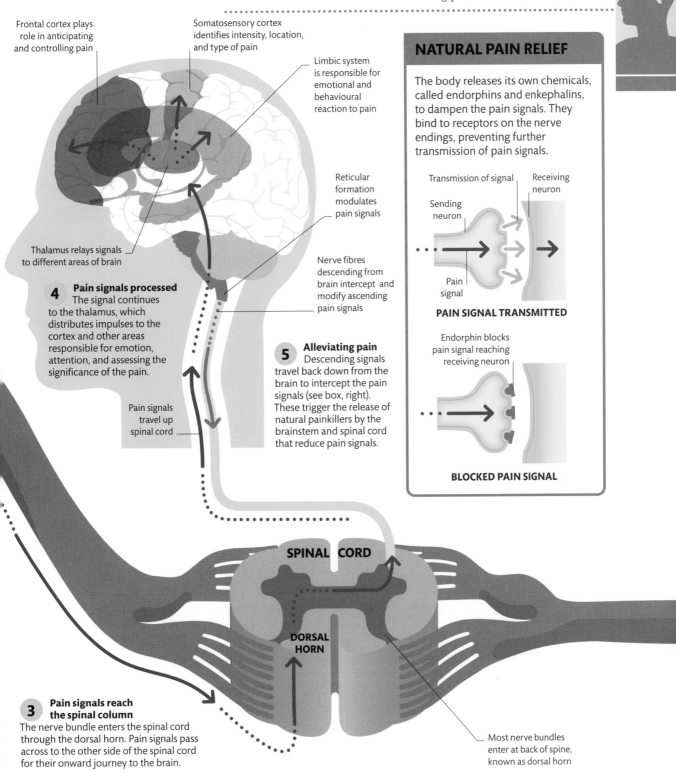

Frontal cortex plays role in anticipating and controlling pain

Somatosensory cortex identifies intensity, location, and type of pain

Limbic system is responsible for emotional and behavioural reaction to pain

Reticular formation modulates pain signals

Thalamus relays signals to different areas of brain

Nerve fibres descending from brain intercept and modify ascending pain signals

4 Pain signals processed
The signal continues to the thalamus, which distributes impulses to the cortex and other areas responsible for emotion, attention, and assessing the significance of the pain.

Pain signals travel up spinal cord

5 Alleviating pain
Descending signals travel back down from the brain to intercept the pain signals (see box, right). These trigger the release of natural painkillers by the brainstem and spinal cord that reduce pain signals.

NATURAL PAIN RELIEF

The body releases its own chemicals, called endorphins and enkephalins, to dampen the pain signals. They bind to receptors on the nerve endings, preventing further transmission of pain signals.

Transmission of signal

Receiving neuron

Sending neuron

Pain signal

PAIN SIGNAL TRANSMITTED

Endorphin blocks pain signal reaching receiving neuron

BLOCKED PAIN SIGNAL

SPINAL CORD

DORSAL HORN

3 Pain signals reach the spinal column
The nerve bundle enters the spinal cord through the dorsal horn. Pain signals pass across to the other side of the spinal cord for their onward journey to the brain.

Most nerve bundles enter at back of spine, known as dorsal horn

How to use your brain to manage pain

When we are in pain, the usual courses of action involve medical treatment or painkillers. However, we can also help control pain ourselves by regulating our mental response – both to the pain and to the stress it causes.

Pain is an emotional as well as a physical response to injury or disease. Intense fear or anxiety are vital immediate reactions that cause you to avoid sources of pain whenever possible. Sometimes, however, pain persists even when the injury or disease is no longer present. A painful sensation can become associated with constant stress, recurring unpleasant memories of what caused the pain, or the constant fear that it will persist or recur.

These feelings can be powerful and unsettling. Although you should always seek medical advice if pain is severe or prolonged, you can also use several techniques to regulate it by training your mind.

The painkiller problem

Medication is often essential to control pain in the short term, but taking painkillers for an extended period can lead to issues such as addiction, or serious physical side-effects, including stomach ulcers and liver disease. Your body may also build up a tolerance to a drug, so that you derive less benefit from it as time goes on.

Mind–body therapies

In addition to medication, you can use mind–body techniques such as relaxation and visualization to reduce or help control pain – with no risk of side-effects. Most use relaxation and deep, controlled breathing to reduce the tension that comes with pain and often makes it worse. Try lying quietly in a darkened room; breathe in deeply while counting to 10, hold the breath for a moment, then exhale slowly for a count of 10. Continue this for 10–20 minutes.

Shifting your attention often reduces pain's severity. Try turning your attention away from the painful area, focusing instead on a non-painful part of your body. Alternatively, imagine the pain as a big ball of energy outside your body, and "shrink" it in your mind. Cognitive behavioural therapy (CBT) uses a similar approach, by training you to replace negative thoughts like "This pain is unbearable", or "I can't stop this pain", with positive ones such as, "This pain is only temporary".

Practising mindfulness reduces stress, making you better able to cope with pain. In this practice, adapted from Buddhist teachings, you merely acknowledge the pain – instead of allowing it to dominate your thoughts or exhausting yourself by actively fighting it.

To sum up, your brain can be a powerful tool for pain control if you:
- **Practise relaxation and deep breathing techniques to reduce stress levels.**
- **Employ mental exercises to shift attention away from pain.**
- **Use CBT techniques to focus on positive thoughts.**
- **Practise mindfulness.**

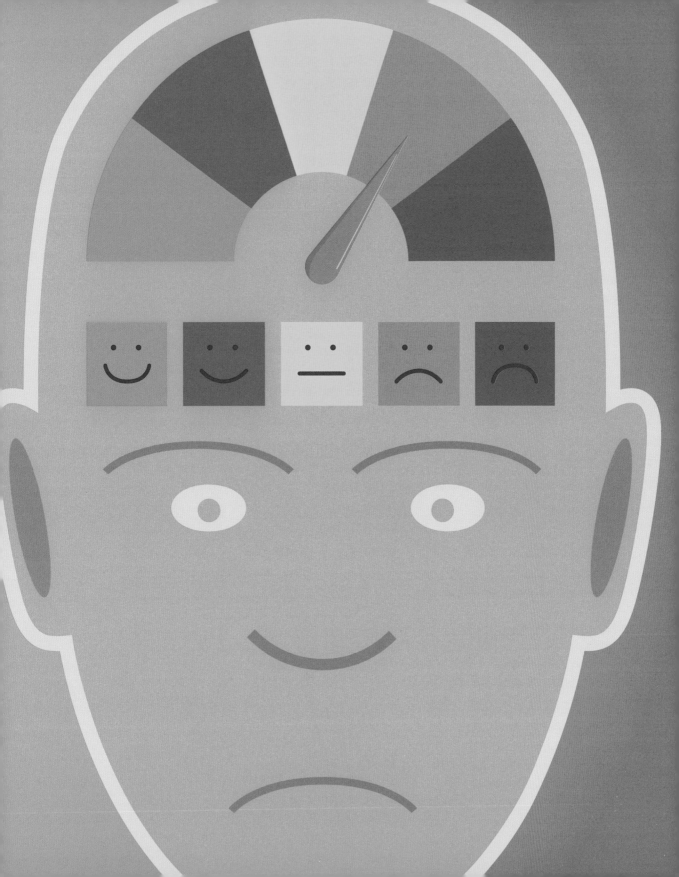

The regulatory system

The human body is a cooperative of 38 trillion cells organized into different systems. Keeping them functioning at their best is a system of feedback mechanisms controlled by the brain.

Maintaining stability

The process of maintaining a stable internal environment is called homeostasis. Key functions, such as breathing, heart rate, pH, temperature, and ion balances have to be kept within strict operating limits to prevent us becoming ill. As the body works, its systems are constantly being moved away from their balance or set point (the value at which a system works best). When the change becomes too great, the body initiates a feedback loop that returns the system to its ideal level. Many of these functions are controlled by a part of the brainstem called the reticular formation.

GENERAL ANAESTHETICS

A vital part of modern surgery, how general anaesthetics work is not fully understood. What is known is that they act on the reticular activating system (comprising the reticular formation and its connections) to suppress awareness, and on the hippocampus to temporarily suspend memory formation. Anaesthetics also affect the nuclei of the thalamus, preventing the flow of sensory information from the body to the brain.

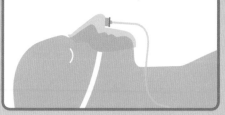

3 Signals forwarded
Signals are then sent directly to the thalamus and hypothalamus, as well as to the appropriate areas of the cerebral cortex for a decision and response to the stimulus.

Signals travel to various areas of cerebral cortex

Excitatory area of reticular formation amplifies important signals

THALAMUS

Hypothalamus regulates sleep, hunger, and body temperature

Thalamus relays sensory signals to cerebral cortex

MEDULLA

2 Signals processed
In the reticular formation, unwanted signals are suppressed in the inhibitory area, while others are amplified in the excitatory area.

Inhibitory area of reticular formation dampens unwanted signals

SPINAL CORD

WHAT IS THE RETICULAR FORMATION?

The reticular formation consists of more than 100 nuclei that project to the forebrain, cerebellum, and brainstem, controlling many of the body's vital functions.

Impulses travel up spinal cord

1 Signals travel up the spinal column
Incoming sensory signals from all over the body travel to the reticular formation.

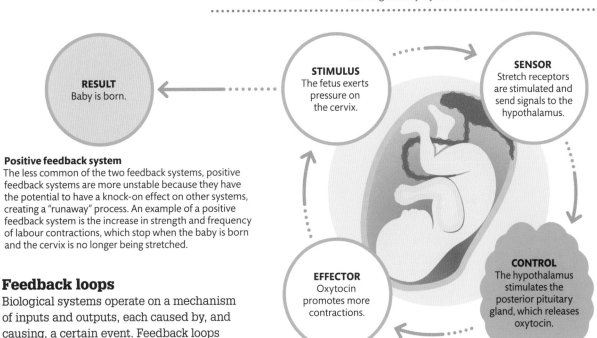

Positive feedback system

The less common of the two feedback systems, positive feedback systems are more unstable because they have the potential to have a knock-on effect on other systems, creating a "runaway" process. An example of a positive feedback system is the increase in strength and frequency of labour contractions, which stop when the baby is born and the cervix is no longer being stretched.

Feedback loops

Biological systems operate on a mechanism of inputs and outputs, each caused by, and causing, a certain event. Feedback loops either amplify the output of a system (positive feedback) or inhibit the output of the system (negative feedback). Feedback loops are important because they allow living organisms to maintain homeostasis.

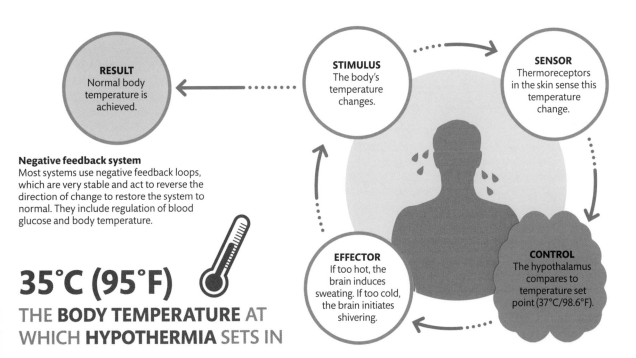

Negative feedback system

Most systems use negative feedback loops, which are very stable and act to reverse the direction of change to restore the system to normal. They include regulation of blood glucose and body temperature.

35°C (95°F)

THE **BODY TEMPERATURE** AT WHICH **HYPOTHERMIA** SETS IN

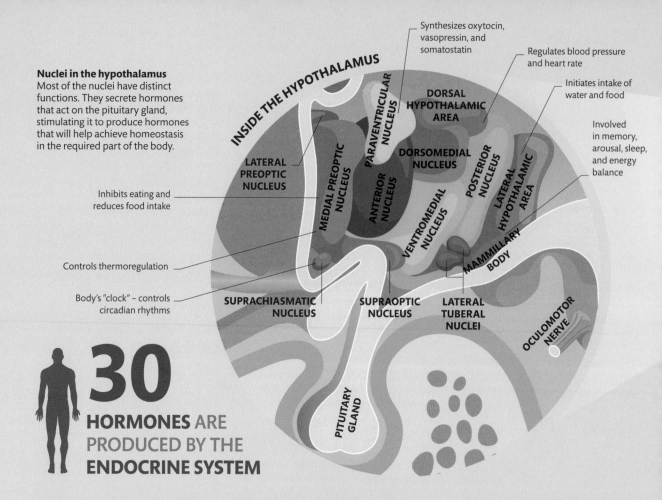

INSIDE THE HYPOTHALAMUS

Synthesizes oxytocin, vasopressin, and somatostatin

Regulates blood pressure and heart rate

Initiates intake of water and food

Involved in memory, arousal, sleep, and energy balance

Nuclei in the hypothalamus
Most of the nuclei have distinct functions. They secrete hormones that act on the pituitary gland, stimulating it to produce hormones that will help achieve homeostasis in the required part of the body.

Inhibits eating and reduces food intake

Controls thermoregulation

Body's "clock" – controls circadian rhythms

DORSAL HYPOTHALAMIC AREA

PARAVENTRICULAR NUCLEUS

DORSOMEDIAL NUCLEUS

LATERAL PREOPTIC NUCLEUS

MEDIAL PREOPTIC NUCLEUS

ANTERIOR NUCLEUS

POSTERIOR NUCLEUS

LATERAL HYPOTHALAMIC AREA

VENTROMEDIAL NUCLEUS

MAMMILLARY BODY

SUPRACHIASMATIC NUCLEUS

SUPRAOPTIC NUCLEUS

LATERAL TUBERAL NUCLEI

OCULOMOTOR NERVE

PITUITARY GLAND

30

HORMONES ARE PRODUCED BY THE **ENDOCRINE SYSTEM**

Neuroendocrine system

Maintaining homeostasis (see p.90) requires the brain and body to communicate. This is achieved using chemical messengers called hormones.

The hypothalamus

At the centre of the brain's homeostasis system is the hypothalamus (see p.34). It contains clusters of neurons, called nuclei, that perform specific functions and has connections to the autonomic nervous system (see p.13), through which it sends messages to control heart rate, digestion, and breathing. When the hypothalamus receives a signal from the nervous system, it secretes neurohormones, which in turn stimulate the pituitary gland to secrete hormones. These affect organs all over the body and prompt them to increase or suppress their own hormone production.

OUT OF BALANCE

When homeostasis is disrupted, it can lead to disease, as well as to our cells malfunctioning. The body tries to correct the problem but may make it worse, depending on what is influencing the imbalance. Genetics, lifestyle, and toxins can all impact homeostasis.

Hormone producers

Hormones are used for two types of communication. The first is between two endocrine glands, where a hormone is released to stimulate a target gland to alter the amount of hormone it is secreting. The second is between a gland and a target organ, such as the release of insulin from the pancreas prompting muscle cells to take up glucose.

Hypothalamus links nervous system to endocrine system

Pineal gland releases melatonin in response to light levels – melatonin governs body's circadian rhythm and regulates some reproductive hormones

Controlled by the hypothalamus, pituitary gland acts as "master gland"; it secretes its own hormones that control other glands

Thyroid gland and parathyroid glands regulate metabolism, blood calcium levels, and heart rate

PARATHYROID GLAND

THYROID GLAND

Produces cortisol (regulates metabolism, immune response, and energy conversion), aldosterone (controls blood pressure and salt balance), and adrenaline (fight or flight hormone)

Produces white blood cells that defend against viruses and infections

THYMUS

Releases hunger-inducing hormone ghrelin and hormone gastrin, which stimulates acid production

STOMACH

Secretes renin and angiotensin, which control blood pressure, as well as erythropoietin, which stimulates production of red blood cells

ADRENAL GLAND

KIDNEY

KIDNEY

Secretes insulin, glucagon, and somatostatin to control blood sugar; gastrin, which stimulates stomach cells to produce acid; and a hormone that controls water secretion and absorption in intestines

PANCREAS

Producing hormones

The endocrine system is made up of glands that are dedicated specifically to secreting hormones, as well as organs – such as the stomach – that are not glands themselves but are able to produce, store, and release hormones. Both types react to signals from the brain by increasing or decreasing the production of hormones, which then travel, via the bloodstream, to a target organ, where they lock onto specialized receptors on the surfaces of cells. This triggers a physiological change that restores homeostasis.

Produces female reproductive hormones oestrogen and progesterone, which prepare uterus for menstruation or pregnancy

OVARY

Produce testosterone, which is essential in sperm production, maintaining muscle mass and strength, libido, and bone density

TESTES

Hunger and thirst

Food and drink are essential to human survival. Prompts by hormones to take in nutrients and water are experienced by the body as hunger and thirst.

Hunger

There are two types of hunger. Hedonic hunger involves eating food – particularly foods high in fat, sugar, and salt – when we are already full, while homeostatic hunger (see right) is a response to our energy stores depleting. Once food has passed through the stomach and intestines, the now-empty stomach releases a hormone called ghrelin. This acts on neurons in the hypothalamus to tell us that we are hungry, prompting us to eat. A hunger-inhibiting hormone called leptin is then released by adipose (fat-bearing) tissue to stop us from overeating.

Feeling hungry

The brain, digestive system, and fat stores form an interconnected system that regulates our feelings of hunger. The sensation of hunger can be caused by internal factors, such as our stomach being empty or our blood sugar levels falling, or by external triggers, such as seeing or smelling food.

DEHYDRATION AFFECTS OUR SHORT-TERM MEMORY, CONCENTRATION, AND ANXIETY LEVELS

5 Feeling full
Signals that leptin and insulin levels are increasing stimulate the hypothalamus to produce the hormone melanocortin, which makes us feel full.

4 Signals from adipose tissue
To prevent us from overeating, adipose tissue cells release a hunger-inhibiting hormone called leptin, which travels to the hypothalamus.

3 Signals from pancreas
After we have eaten, the small intestine releases the hormone incretin. This, combined with the stomach stretching and increased glucose in the blood, causes the pancreas to release insulin.

2 Urge to eat
Rising levels of ghrelin instruct the hypothalamus to release a chemical signal called neuropeptide Y, which stimulates our appetite.

Incretin produced by intestines triggers insulin production

KEY

- ⋯▶ Ghrelin
- ⋯▶ Insulin
- ⋯▶ Leptin
- ⋯▶ Incretin
- ⋯▶ Vagus nerve signal
- ⋯▶ Movement of food

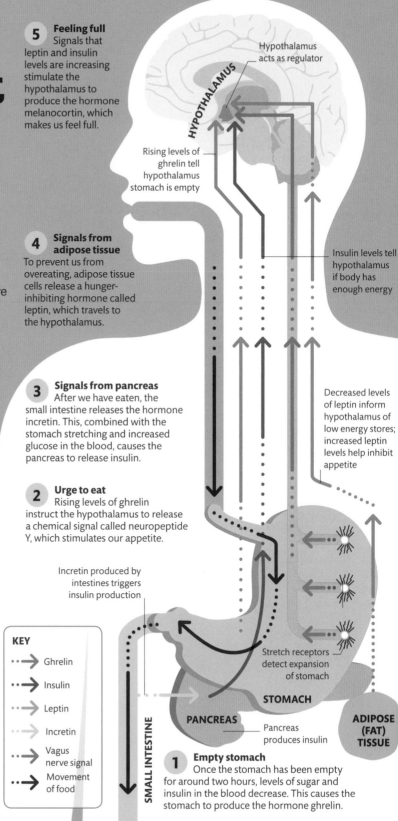

Hypothalamus acts as regulator

HYPOTHALAMUS

Rising levels of ghrelin tell hypothalamus stomach is empty

Insulin levels tell hypothalamus if body has enough energy

Decreased levels of leptin inform hypothalamus of low energy stores; increased leptin levels help inhibit appetite

Stretch receptors detect expansion of stomach

STOMACH

PANCREAS

Pancreas produces insulin

SMALL INTESTINE

ADIPOSE (FAT) TISSUE

1 Empty stomach
Once the stomach has been empty for around two hours, levels of sugar and insulin in the blood decrease. This causes the stomach to produce the hormone ghrelin.

Thirst

When water levels in the body drop, salt levels in the blood increase. Thirst areas in the brain detect rising salt levels and signal to the body to increase water levels by reducing urine output and taking in more fluids. After drinking, it takes around 15 minutes before salt concentration levels in the blood return to normal. It is thought that the gulping action of the throat when swallowing liquids sends signals to stop drinking.

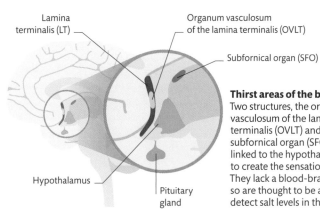

Lamina terminalis (LT)
Organum vasculosum of the lamina terminalis (OVLT)
Subfornical organ (SFO)
Hypothalamus
Pituitary gland

Thirst areas of the brain
Two structures, the organum vasculosum of the lamina terminalis (OVLT) and the subfornical organ (SFO) – both linked to the hypothalamus – help to create the sensation of thirst. They lack a blood-brain barrier, so are thought to be able to detect salt levels in the blood.

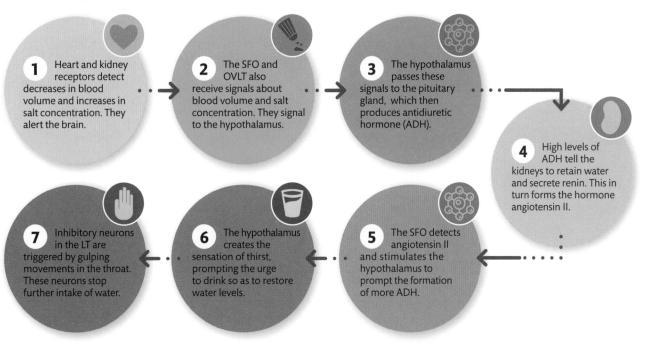

1 Heart and kidney receptors detect decreases in blood volume and increases in salt concentration. They alert the brain.

2 The SFO and OVLT also receive signals about blood volume and salt concentration. They signal to the hypothalamus.

3 The hypothalamus passes these signals to the pituitary gland, which then produces antidiuretic hormone (ADH).

4 High levels of ADH tell the kidneys to retain water and secrete renin. This in turn forms the hormone angiotensin II.

5 The SFO detects angiotensin II and stimulates the hypothalamus to prompt the formation of more ADH.

6 The hypothalamus creates the sensation of thirst, prompting the urge to drink so as to restore water levels.

7 Inhibitory neurons in the LT are triggered by gulping movements in the throat. These neurons stop further intake of water.

HOW LONG CAN YOU SURVIVE WITHOUT FOOD OR WATER?

Three to four days is the average without water, but you can go up to two months without food in certain circumstances.

ARE YOU DEHYDRATED?

The most obvious symptoms of dehydration are a dry mouth and eyes, and perhaps a slight headache. Another good way to tell is by the colour of your urine. It should be pale yellow at full hydration. A darker amber colour shows severe dehydration. Adults should take in around 2–2.5 litres (3.5–4 pints) of fluids a day.

VERY HYDRATED
HYDRATED
MODERATELY DEHYDRATED
VERY DEHYDRATED
DANGEROUSLY DEHYDRATED

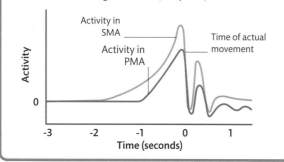
THE **CEREBELLUM** CONTAINS MORE THAN **50 PER CENT** OF THE BRAIN'S **NEURONS**

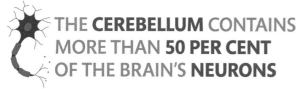

Putamen feeds stored information to posterior parietal cortex

Posterior parietal cortex receives information from putamen and also assesses body's position in relation to surroundings

DORSOLATERAL FRONTAL CORTEX

POSTERIOR PARIETAL CORTEX

BASAL GANGLIA

PUTAMEN

THALAMUS

VISUAL CORTEX

SPINAL CORD

Sensory information is sent from visual cortex via thalamus to dorsolateral frontal cortex

1 Gathering information
Sensory areas, such as the visual cortex, send signals to the frontal cortex. The putamen, which stores learned actions, sends information to the parietal cortex, which assesses whether these learned actions could be used in this new situation.

Planning movement

Conscious movements are those that we deliberately decide to make. They involve several regions of our brain and include processes that lie outside our conscious awareness.

The planning process

There are several stages involved in carrying out a movement – from initial perception of the environment, to planning, to adjustments during the movement. These stages involve different areas of the brain working together to produce a response. The area that prompts the movement is the motor cortex. Different sections of the motor cortex send signals to different parts of the body (see p.98). However, before an action begins an action plan is created by the dorsolateral frontal cortex and the posterior parietal cortex, and is passed through two areas of the motor cortex: the supplementary motor area (SMA) and the premotor area (PMA). The cerebellum coordinates the movement as it is happening. The steps above show the brain areas involved and the sequence of signals in a typical movement.

WHY DON'T WE FORGET HOW TO RIDE A BICYCLE?

Nerve cells in the putamen encode the sequence of muscle movements into our long-term memory storage so that they are easily accessible even years later.

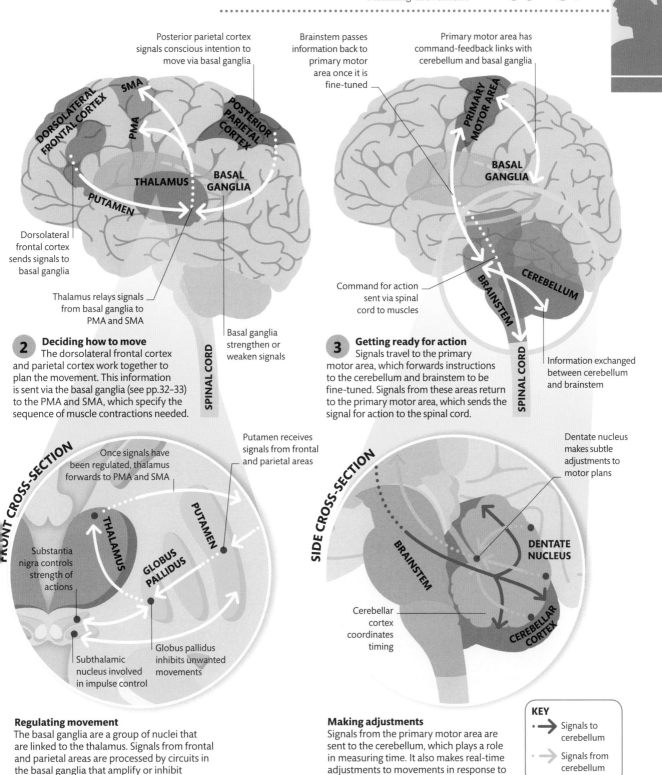

Posterior parietal cortex signals conscious intention to move via basal ganglia

Brainstem passes information back to primary motor area once it is fine-tuned

Primary motor area has command-feedback links with cerebellum and basal ganglia

SMA

DORSOLATERAL FRONTAL CORTEX

PMA

POSTERIOR PARIETAL CORTEX

PRIMARY MOTOR AREA

THALAMUS

BASAL GANGLIA

BASAL GANGLIA

PUTAMEN

Dorsolateral frontal cortex sends signals to basal ganglia

Thalamus relays signals from basal ganglia to PMA and SMA

Basal ganglia strengthen or weaken signals

Command for action sent via spinal cord to muscles

BRAINSTEM

CEREBELLUM

SPINAL CORD

SPINAL CORD

Information exchanged between cerebellum and brainstem

2 Deciding how to move
The dorsolateral frontal cortex and parietal cortex work together to plan the movement. This information is sent via the basal ganglia (see pp.32–33) to the PMA and SMA, which specify the sequence of muscle contractions needed.

3 Getting ready for action
Signals travel to the primary motor area, which forwards instructions to the cerebellum and brainstem to be fine-tuned. Signals from these areas return to the primary motor area, which sends the signal for action to the spinal cord.

FRONT CROSS-SECTION

Once signals have been regulated, thalamus forwards to PMA and SMA

Putamen receives signals from frontal and parietal areas

THALAMUS

PUTAMEN

Substantia nigra controls strength of actions

GLOBUS PALLIDUS

Subthalamic nucleus involved in impulse control

Globus pallidus inhibits unwanted movements

Dentate nucleus makes subtle adjustments to motor plans

SIDE CROSS-SECTION

BRAINSTEM

DENTATE NUCLEUS

Cerebellar cortex coordinates timing

CEREBELLAR CORTEX

Regulating movement
The basal ganglia are a group of nuclei that are linked to the thalamus. Signals from frontal and parietal areas are processed by circuits in the basal ganglia that amplify or inhibit movement signals.

Making adjustments
Signals from the primary motor area are sent to the cerebellum, which plays a role in measuring time. It also makes real-time adjustments to movements in response to our environment.

KEY
→ Signals to cerebellum
→ Signals from cerebellum

Making a move

Once our brain has planned a movement (see pp.96–97), it sends signals to the appropriate muscles in the body, via the nervous system, to turn intention into action.

From brain to spine

Signals from the motor and parietal areas of the cortex are sent along the axons of neurons, through the brainstem, to communicate with motor neurons in the spinal cord. Most of the axons form part of a bundle called the lateral corticospinal tract, which crosses over at the base of the brainstem so that axons from one brain hemisphere connect to motor nerves for the opposite side of the body. Other nerve tracts originate in different parts of the midbrain and perform specific movement functions.

SIMPLE AND COMPLEX MOVEMENTS

A motor homunculus shows which areas of the motor cortex control which areas of the body. Areas for adjacent body parts – such as the arm and hand – are generally grouped together. The body parts are shown in proportion; those areas that make complex movements, such as the face and the hand, take up more space in the cortex than those making simple movements, such as the foot.

PRIMARY MOTOR AREA

MOTOR HOMUNCULUS

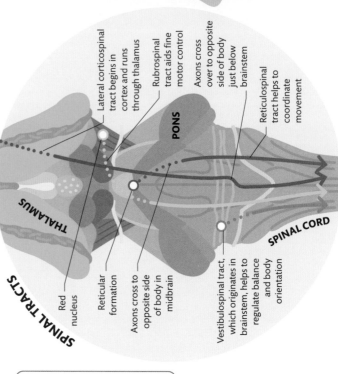

LEFT SIDE OF BRAIN

PARIETAL CORTEX

PRIMARY MOTOR AREA

Most signals originate in primary motor area

MIDBRAIN

CEREBELLUM

SPINAL CORD

Axons collect in midbrain and join spinal cord

Neurons from brain (upper motor neurons) pass signals down spinal cord

SPINAL TRACTS

THALAMUS

Red nucleus

Reticular formation

Axons cross to opposite side of body in midbrain

Vestibulospinal tract, which originates in brainstem, helps to regulate balance and body orientation

Lateral corticospinal tract begins in cortex and runs through thalamus

Rubrospinal tract aids fine motor control

Axons cross over to opposite side of body just below brainstem

Reticulospinal tract helps to coordinate movement

PONS

SPINAL CORD

KEY

↑ Lateral corticospinal tract
↑ Rubrospinal tract
↑ Vestibulospinal tract
↑ Reticulospinal tract
↑ Motor-nerve axon

1 Nerve tracts
The axons of the lateral corticospinal tract send signals to muscles that connect to the skeleton to produce voluntary limb movements. Other groups of axons are responsible for the body's involuntary responses, such as balance, as well as for fine-tuning movements.

SPINAL CORD

Upper motor neurons

WHITE MATTER

GREY MATTER

VENTRAL HORN

Lower motor neurons

Lower motor neurons pass signals from spinal cord to muscles

2 The upper and lower motor neurons meet in the ventral horn of the spinal cord. The outer part of the ventral horn carries nerves that run to the hands and feet; the central part carries nerves to the upper arms and thighs.

HOW LONG DOES IT TAKE FOR A SIGNAL TO TRAVEL FROM BRAIN TO MUSCLE?

Signals can travel from the brain to our muscles at a speed of up to 120 m (395 ft) per second.

RADIAL NERVE

MUSCLE

RIGHT ARM

Muscle contracts and moves joint, causing arm to bend

NEUROMUSCULAR JUNCTION

Direction of signal

Acetylcholine

MUSCLE FIBRE

SYNAPTIC CLEFT

AXON TERMINAL

Receptor for acetylcholine

3 At the neuromuscular junction, the end of the axon releases acetylcholine, a neurotransmitter (see p.24). The acetylcholine binds to receptors in the muscle cell membrane. This triggers chemical reactions that make the muscle fibre contract.

Executing movement
Nerve signals make a muscle contract and pull on the associated joint to move the part of the limb just beyond it. Muscles used in fine movements have more nerve endings than those used for simple movements.

From spine to muscle

Inside the spinal cord, the axons of the corticospinal tract, which are covered with a myelin sheath, form the white matter. The grey matter at the centre of the spinal cord consists of the cell bodies of motor neurons. The ends of the corticospinal axons (known as upper motor neurons) synapse on to motor neurons (known as lower motor neurons) in the ventral horn of the grey matter. The axons of the lower neurons exit the spine through gaps in the vertebrae (see p.12), and extend to the muscle fibres. The point where the nerve endings activate the muscle fibres to complete the movement is called the neuromuscular junction.

Unconscious movement

We perform many voluntary actions without having to think about them because they are so familiar. Another kind of unconscious movement is the reflex action – an instinctive response to danger.

Reaction pathways

Visual information is vital in helping us to plan our movements. Information from the visual cortex follows two routes in the brain (see pp.70–71). The upper (or dorsal) route, which leads to the parietal lobe, guides our actions in real time. Meanwhile the lower (or ventral) route, which ends at the temporal lobe, triggers stored visual experiences to help us interpret what we see and respond accordingly.

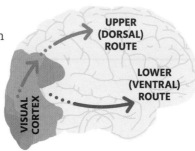

UPPER (DORSAL) ROUTE

LOWER (VENTRAL) ROUTE

VISUAL CORTEX

Visual pathways in the brain
The dorsal route carries information on the position of the body and other objects, while the ventral route draws on perception and memory for identifying objects. The brain uses this information to judge the strength and direction required for a movement.

Coordinated actions

Any sequence of actions demands coordination between different parts of the brain – first to focus attention on the task, then to integrate sensory information and memory to create a plan, then to engage the motor area to act. Acquiring a new skill, such as driving or playing a sport, involves learning and practising movement sequences so that they become almost unconscious. When we learn a skill, our brain cells form new connections. By the time we have mastered a skill (see box, right), there is far less cortical activity associated with performing that task than there was when we were a novice. As a result, the actions of a skilled person – such as a professional tennis player – are more rapid, precise, and subtle.

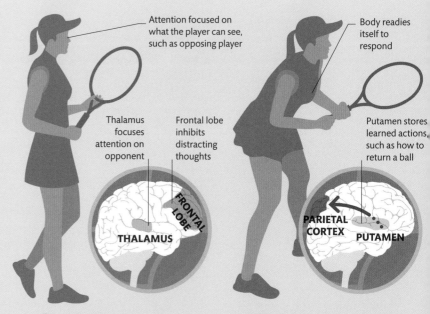

Attention focused on what the player can see, such as opposing player

Body readies itself to respond

Thalamus focuses attention on opponent

Frontal lobe inhibits distracting thoughts

Putamen stores learned actions, such as how to return a ball

FRONTAL LOBE

THALAMUS

PARIETAL CORTEX

PUTAMEN

1 Attention
To prepare for action, the thalamus directs attention to the area where the activity will occur (such as the opposing player), while the frontal lobes block distracting thoughts so the player can concentrate on the visual cues.

2 Memory
Visual cues trigger the parietal cortex to call up memories of action sequences from the putamen. The parietal cortex uses this information to assess the context and create an internal model for the action.

Reflex actions

Reflexes are split-second responses to danger that we do not have to learn or even think about; the body reacts automatically. Reflex actions involve the same muscles that are used in voluntary movements, but the initial, instantaneous response does not involve the brain. Instead, the signal from the sensory nerves travels to the spinal cord, which triggers a response that travels along the motor nerves. Additional signals are sent to the brain afterwards, to encode the memory in case the danger recurs.

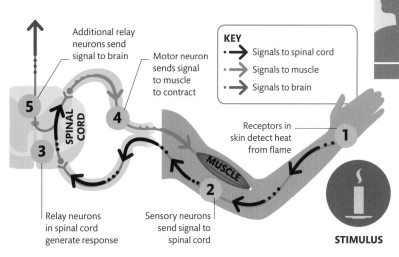

Additional relay neurons send signal to brain

Motor neuron sends signal to muscle to contract

KEY
• → Signals to spinal cord
• → Signals to muscle
• → Signals to brain

Receptors in skin detect heat from flame

5

SPINAL CORD

4

3

MUSCLE

2

1

Relay neurons in spinal cord generate response

Sensory neurons send signal to spinal cord

STIMULUS

Bypassing the brain

Reflexes involve a simple neural response called the reflex arc. Receptors in the skin and muscles send a danger signal along sensory neurons to the spinal cord; there, relay neurons synapse with motor neurons to trigger a fast response.

OUR **NEURONS AND NERVE PATHWAYS CHANGE** CONSTANTLY IN **RESPONSE TO EXPERIENCES**

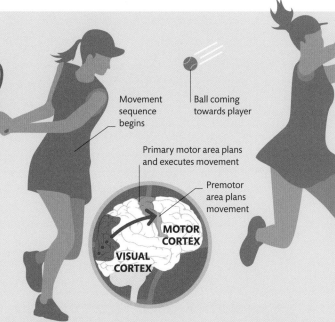

Movement sequence begins

Ball coming towards player

Primary motor area plans and executes movement

Premotor area plans movement

MOTOR CORTEX

VISUAL CORTEX

3 Planning
The brain combines real-time visual information and stored programmes for movement sequences to create a plan of action. This is first rehearsed in the premotor area and then sent to the primary motor cortex.

4 Conscious action
By the time the player becomes conscious of acting, the movement sequence is well underway. The action is most likely to be effective if the person has sufficient skill, stored knowledge, and information.

DEVELOPING COMPETENCE

Anyone learning a new skill passes through several stages. Beginners have to work hard to acquire competence. With practice, neural pathways develop until the learner can perform well without thinking about it.

Unconscious competence
Performing skill is automatic

Conscious competence
Able to use skill, but only with effort

Conscious incompetence
Aware of skill needed but lacking proficiency

Unconscious incompetence
Unaware of skill needed and lack of proficiency

Mirror neurons

Learning does not just involve practising a new skill – we also learn by watching others. This kind of learning is thought to involve nerve cells in the brain called mirror neurons that allow us to experience actions without actually performing them.

What are mirror neurons?

Mirror neurons are brain cells that fire both when we perform an action and when we see someone else performing that action. They were first discovered in monkeys, but have since been found in humans, too. Most studies have used functional magnetic resonance imaging (fMRI, see p.43), but one study involved people who had electrodes implanted into their brains. In this instance, mirror neuron cells were detected in the supplementary motor area, where movement sequences are planned, as well as in the hippocampus, which governs memory and navigation.

Where are they?

Mirror neurons have been found in several cortical areas, as well as in structures deeper within the brain, such as the hippocampus.

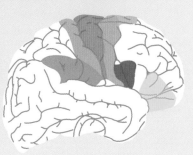

KEY

- 🔵 Premotor area
- ⚫ Part of Broca's area
- ⚪ Inferior frontal gyrus
- 🔵 Supplementary motor area
- 🔴 Primary motor area
- 🔵 Somatosensory area
- 🔵 Inferior parietal area

Mirroring movement

Some scientists suggest that mirror neurons may play a role in learning how to imitate movement. In this theory, information on the purpose of an action is passed to mirror neurons from brain areas such as the prefrontal cortex, which is responsible for analysis. Mirror neurons in various motor areas then encode a simulation of that action, which becomes part of our own motor programming. We can then go on to use this "program" if we need to carry out the action ourselves.

Observing an action

Mirror neurons respond differently to various actions of the face and limbs. In particular, neurons in different brain areas are activated for movements of the body itself, such as chewing, and those focused on a visible object, such as biting a fruit.

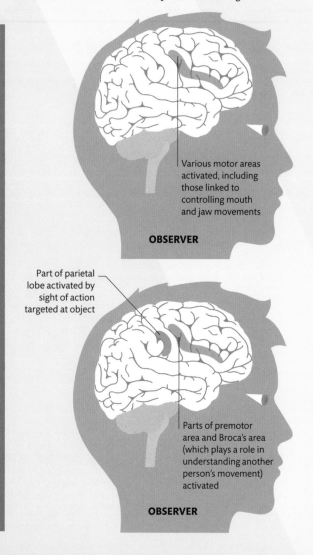

Various motor areas activated, including those linked to controlling mouth and jaw movements

OBSERVER

Part of parietal lobe activated by sight of action targeted at object

Parts of premotor area and Broca's area (which plays a role in understanding another person's movement) activated

OBSERVER

DO OTHER ANIMALS HAVE MIRROR NEURONS?

Mirror neurons were first discovered in macaque monkeys. They have also been found in some birds, such as songbirds, and more recently in rats.

YAWNING

Mirror neurons may play a role in "contagious yawning" – the impulse to yawn when we see someone else yawning. FMRI scans of people who watched videos of someone else yawning showed activity in the right inferior frontal gyrus, an area associated with mirror neurons.

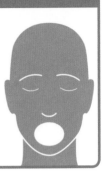

1 Watching a body movement
Watching a person perform an action not linked to an object, such as chewing, activates the premotor area in the observer. This is an area that is linked to rehearsing planned sequences of action. It also activates areas in the primary motor area associated with mouth and jaw movements.

ACTION WITH NO OBJECT

Understanding intention

Mirror neurons are activated in different ways when we see others performing particular actions, suggesting they could play a role in decoding intention. Watching similar actions observed in different contexts – such as watching someone pick up a cup either to drink from it or to tidy it up – triggers different levels of neural activity in the inferior frontal gyrus; an area of the brain that directs our attention to objects in our environment.

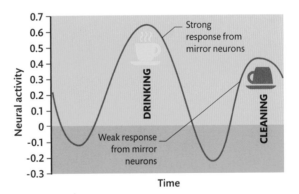

Intention and brain activity
Activity in the brain is greater when a person watches someone lift a cup to drink rather than when they watch someone pick it up to clear it away. Some scientists suggest this may be because drinking has a greater biological function than cleaning.

2 Watching action on an object
Watching an action directed at an object, such as a person biting into a fruit, activates similar areas of the motor cortex. However, mirror neurons also fire in an additional area, the parietal cortex, which is involved in interpreting sensory input as well as providing information about the body's position.

ACTION WITH AN OBJECT

THE **BRAINWAVES** OF **MUSICIANS** COME INTO **SYNC** WHEN THEY **PLAY TOGETHER**

COMMUNICATION

Emotions

Emotions are physiological responses to external events, shaped by experience, that are accompanied by distinctive feelings. They evolved to push us away from danger and towards reward.

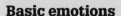

HORMONES THAT TRIGGER **EMOTIONAL** RESPONSES ARE ABSORBED IN **6 SECONDS**

Basic emotions

Research suggests that there are four physiologically distinct conscious feelings: anger, fear, happiness, and sadness. Aspects of these combine and allow us to feel a range of emotions. Broadly, emotions are positive or negative experiences, which vary in intensity. Different emotional states are associated with particular physiological changes that affect how a person behaves and thinks. For example, we view the world differently when we are relaxed and when we are afraid. This coordination of physiology, behaviour, and thought with feeling is what makes us adapt our behaviour in response to events.

Emotions
Other emotional experiences stem from the four key ones. A recent study found there may be 27 types of emotional experience, some of which are shown here. Certain emotions lie along gradients, such as moving from anxiety to fear to horror.

WHY DO WE CRY?

Only humans cry, and nobody is certain why we do it – especially given that both sadness and joy can evoke tears. Crying serves an interpersonal function, signalling that we are in emotional distress to evoke appropriate social responses. It is also cathartic, enabling full emotional engagement and processing that is good for mental health.

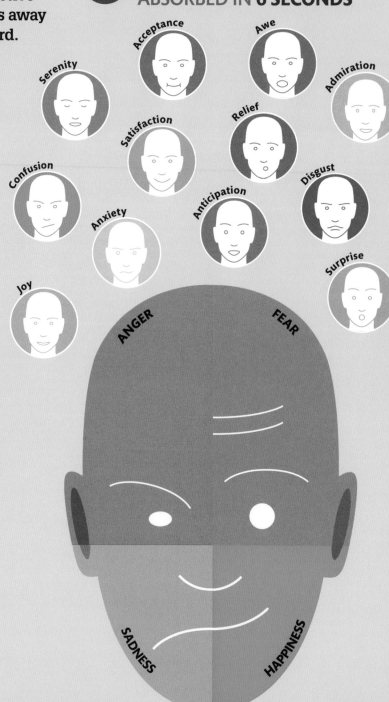

The anatomy of emotion

In response to a stimulus, the brain initiates hormonal changes that, in turn, trigger physiological changes that prime us to respond in appropriate ways to the current emotional state. Heart rate changes, altered blood flow to the muscles, and sweating are associated with heightened emotions. These changes can be felt consciously, increasing the emotional intensity.

WHAT IS THE PURPOSE OF LAUGHTER?

The relaxation that results from a bout of laughter inhibits the biological fight or flight response.

Happiness and sadness
Serotonin, dopamine, oxytocin, and endorphins are hormones that affect our happiness profoundly. Emotions are felt across the body, with different emotions felt in different places. The effects of serotonin are shown here.

SEROTONIN

Brain produces majority of hormones relating to happiness

Heart rate decreases

Large amount of serotonin produced by large intestine

Feeling of wellbeing reported all over body

HAPPINESS

Low levels of serotonin in brain

Mildly increased bodily feelings around neck and chest

Low levels of serotonin produced

Sensations of decreased limb activity

SADNESS

KEY
○ Positive feelings reported
● Negative feelings reported

Unconscious emotions

For primitive automatic responses, such as the fight-or-flight reflex, speed is critical. Emotionally charged stimuli presented too fast to be consciously perceived can evoke emotional responses and activate the amygdala. These initial responses shape how the cortex processes the information. The amygdala is involved in emotional memory that may be automatically activated in future.

SLOW AND ACCURATE ROUTE

Sensory cortex
Sensory information transmitted to the sensory cortex is extensively processed towards conscious perception and integrated with stored information. This takes time.

Hippocampus
The hippocampus processes consciously perceived information to form memories. It also compares incoming signals to previous memories to adjust emotional responses.

Thalamus
Incoming information is relayed to the amygdala for quick assessment and action, and also to cortical areas where it enters conscious awareness.

QUICK AND DIRTY ROUTE

Amygdala
The amygdala instantly assesses the emotional importance of incoming information content and rapidly sends signals to other areas for immediate bodily action.

Hypothalamus
Signals from the amygdala trigger hormonal changes and output to the autonomic nervous system to prime the body to respond to emotional stimuli.

Two routes
Conscious processing of emotions involves integrating sensory information with stored memories and reasoned evaluations of a situation – this is the "slow and accurate route". In contrast, unconscious responses, via the "quick and dirty route", happen much faster. The prefrontal cortex is important in conscious emotional regulation.

Fear and anger

Fear and anger trigger the release of hormones in the body that prepare us to deal with threats. In the modern world, however, long-term anxiety can cause over-activation of the sympathetic nervous system and lead to health problems.

Fight or flight

When we see a possible threat, visual information travels to our amygdala, a tiny part of the brain that processes emotion. The amygdala sends a signal to the hypothalamus, which activates the sympathetic nervous system, preparing the body to react to danger (see p.13). The hypothalamus also sends signals to the pituitary and adrenal glands, which secrete hormones such as cortisol and adrenaline. The combined effect of these pathways is to initiate our fight-or-flight reflex, which prepares our bodies to attack or run away.

Responding to danger
Signals travel to the thalamus and amygdala, which triggers the hypothalamus to produce fight or flight hormones. A slower, conscious pathway involving the cortex also assesses the situation (see p.107).

Pupils dilate
Our pupils enlarge, letting in more light so we can see the threat more clearly.

Blood vessels constrict
Blood flow is directed away from the surface of the skin, so we may appear pale.

Sweating increases
Our sweat glands are triggered, and we begin to sweat, ensuring we remain cool if physical exertion is needed.

Heart rate increases
Our heart beats faster to pump oxygen-and-nutrient-rich blood to where it is needed in the body.

Saliva production reduces
Saliva secretion slows down when we are afraid. This causes a dry mouth.

Breathing rate rises
This oxygenates our muscles, preparing them for action. But it can also cause symptoms of hyperventilation.

Digestion slows
To avoid wasting energy, digestive activity falls. In extreme cases, we may vomit to eject undigested food.

Muscles tense
The muscles in our arms, legs, and shoulders prepare themselves for action. We may feel tense or "wound up".

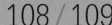

4 PER CENT OF PEOPLE WORLDWIDE HAVE ARACHNOPHOBIA, A FEAR OF SPIDERS

Immune system activity reduced
In the moment, dealing with infections is not crucial, so the immune system shuts down to save energy.

Blood sugar spikes
Sugar stores are released from the liver, to provide the muscles with the energy they need to work. Fat stores are also mobilized.

Blood flows to muscles
Blood carries nutrients and oxygen to the muscles, readying them to fight or flee from danger.

Bladder muscles relax
This causes us to need to urinate, which rids the body of excess weight and makes us faster and lighter.

Panic attacks

Panic attacks are physical reactions to fear or anxiety. Symptoms include a pounding heart, chest pain, rapid, shallow breathing, and sweating. Initially, sufferers can think they are having a heart attack. The first step to break the cycle is to recognize that you are experiencing a panic attack.

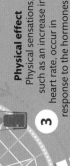
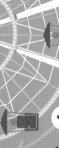
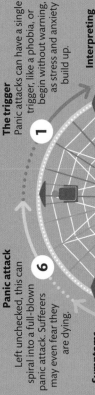

The panic cycle

1 The trigger
Panic attacks can have a single trigger, like a phobia, or begin without warning, as stress and anxiety build up.

2 Interpreting danger
Your brain construes the feelings as danger and releases fight-or-flight hormones.

3 Physical effect
Physical sensations, such as an increase in heart rate, occur in response to the hormones.

4 Anxiety builds
Unaware of the triggers and unsure why this is happening, your anxiety increases.

5 Symptoms increase
More hormones are released, and symptoms get worse, increasing anxiety further.

6 Panic attack
Left unchecked, this can spiral into a full-blown panic attack. Sufferers may even fear they are dying.

Angry or afraid?

The bodily reactions to fear and anger are similar. It is mainly the way we interpret the sensations we experience that determines whether we feel afraid or angry. One theory suggests that if we know why a negative event happened, and who was responsible for it, we will feel angry. If we are unable to work out the cause, or it is out of our control, we will feel fear.

Context is key
Whether we react with fear or anger to a particular stimulus is often conditioned by its context.

You are woken by loud noises downstairs in the middle of the night.

Fight-or-flight reflex is triggered

YOU LIVE ALONE
You live alone, so know there should not be anyone here.

Without being able to work out the cause, you feel afraid.

LIVE WITH FLATMATE
You recall your flatmate was out, and realize she has come back.

Sensations are interpreted as anger about inconsiderate behaviour.

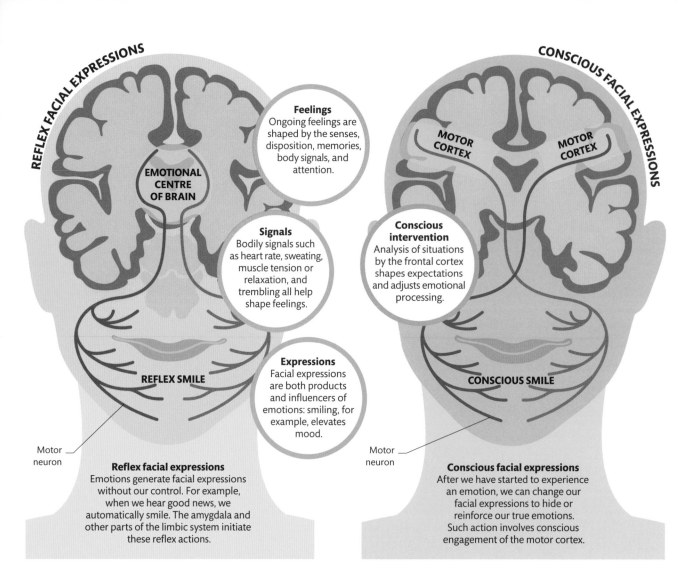

Feelings
Ongoing feelings are shaped by the senses, disposition, memories, body signals, and attention.

EMOTIONAL CENTRE OF BRAIN

Signals
Bodily signals such as heart rate, sweating, muscle tension or relaxation, and trembling all help shape feelings.

MOTOR CORTEX

MOTOR CORTEX

Conscious intervention
Analysis of situations by the frontal cortex shapes expectations and adjusts emotional processing.

Expressions
Facial expressions are both products and influencers of emotions: smiling, for example, elevates mood.

REFLEX SMILE

CONSCIOUS SMILE

Motor neuron

Motor neuron

Reflex facial expressions
Emotions generate facial expressions without our control. For example, when we hear good news, we automatically smile. The amygdala and other parts of the limbic system initiate these reflex actions.

Conscious facial expressions
After we have started to experience an emotion, we can change our facial expressions to hide or reinforce our true emotions. Such action involves conscious engagement of the motor cortex.

Conscious emotion

Emotions are felt consciously, and whether they are positive or negative, changeable or constant, they have major effects on our quality of life. Conscious feelings ceaselessly interact with the unconscious processes that also shape our emotions.

How emotions form
Both reflex and conscious expressions are mediated by the motor cortex, but reflex ones are signalled to the motor area directly from the limbic system rather than via the frontal lobes. We can also consciously modify our physical responses to emotion.

Forming emotions

Emotional responses are complex and dynamic. They arise when rapid innate responses to stimuli interact with detailed analysis. Innate responses evolved as the most beneficial reactions to key stimuli. Once such stimuli have caught a person's attention, reasoned assessment follows. Then, how a person's emotions change is shaped by their disposition, past experience, and how they assess multiple streams of information.

Emotional reactions

Emotional responses evolve over time, from initial protective responses to more considered responses. Imagine a friend leaping out on you: first you feel shock or fear, but as the brain processes what is happening, you transition to calm. The first stage involves attention being grabbed and the amygdala responding early to prime the conscious brain to "expect" an important perception.

KEY
- Amygdala
- Primary visual cortex
- Frontal cortex
- Fusiform gyrus (face recognition area)
- Motor cortex
- Parietal cortex

Less than 100 milliseconds
Sensory information goes to the amygdala, which sends signals to the parietal cortex and then to the motor cortex to produce fast reactions to emotional stimulus, such as when fleeing from danger.

100–200 milliseconds
The information then arrives in the frontal lobes, where it becomes conscious and appropriate action is planned.

Information registers in frontal cortex

350 milliseconds
Considered reactions are then conveyed back to the motor cortex, which signals appropriate bodily responses.

Signal travels to motor and parietal cortex

Signal travels to amygdala

Signal from sensory areas

Recognition pathway

Signal from frontal lobe to motor cortex

SEROTONIN

Alongside dopamine and noradrenaline, serotonin is a neurotransmitter that plays a key role in regulating mood. Although it is not as simple as high serotonin equals happiness and low equals sad, decreases in this molecule are associated with depression and anxiety. Many antidepressant medications act by increasing brain levels of serotonin. Exercise can help, too – for example, taking a brisk walk or dancing can raise serotonin levels.

EMOTIONS ARE **CONTAGIOUS** – HUMANS **MIMIC EACH OTHER'S EXPRESSIONS**

Emotions versus moods

Emotions are usually transient – arising from thoughts, activities, or events that act as cues for adaptive behaviours. Moods last for hours, days, or even months. For example, emotion might be experiencing a sudden rush of joy at seeing a friend waiting to greet you compared to a lingering mood of sadness or worry after losing a job. Emotions tend to be expressed in the moment, while moods are not.

ADAPTIVE BEHAVIOURS		
EMOTION	**POSSIBLE STIMULUS**	**ADAPTIVE BEHAVIOUR**
Anger	Challenging behaviour from another person	"Fight" reaction prompts dominant and threatening stance or action
Fear	Threat from stronger or dominant person	"Flight" to avoid threat; or act to socially appease the threatening person
Sadness	Loss of loved one	Backward-looking state of mind and passivity, to avoid additional challenge
Disgust	Unwholesome object (e.g. rotting food or unclean surroundings)	Aversion behaviour – remove oneself from the unhealthy environment
Surprise	Novel or unexpected event	Attention on object of surprise maximizes sensory input that guides reaction

Reward centres

The brain's reward system evolved because it helped us seek out things that are important for our survival. But if this system is hijacked, it can lead to addiction.

Reward pathways

When we do something that is important for our survival, such as eating when hungry, or having sex, neurons that trigger the release of the neurotransmitter dopamine are activated in the ventral tegmental area (VTA). These send signals to an area called the nucleus accumbens – a rush in dopamine here tells the brain this is a behaviour that should be repeated. Neurons also send signals to the frontal cortex, which focuses attention on the beneficial activity.

Rush of dopamine tells brain to repeat activity

Attention focused on activity

Dopamine neurons activated and project to other brain areas

FRONTAL CORTEX

NUCLEUS ACCUMBENS

SUBSTANTIA NIGRA

VTA

LIMBIC SYSTEM

LIGHT ENTERS EYE

Sensory information registers in limbic system

Route to reward
The reward system starts in the VTA in the midbrain, passes to the nucleus accumbens in the basal ganglia, and then the frontal cortex. Dopamine also travels from the substantia nigra to the basal ganglia. This pathway affects motor control.

1 Stimulus
The initial stimulus can originate outside the body, such as the sight of food, or from within, such as falling glucose levels.

2 Urge
Dopamine released from the VTA to the nucleus accumbens drives us to seek out and work for the reward that is linked to the stimulus.

3 Desire
The urge may be registered as a conscious desire in the cortex, but sometimes it goes undetected, or even opposes our conscious desires.

5 Reward
The reward triggers parts of the brain known as "hedonic hotspots" to release opioid-like neurotransmitters, giving a sense of pleasure.

6 Learning
If the reward is better than expected, the brain releases more dopamine, strengthening the connection between stimulus and reward.

4 Action
A region of the frontal cortex weighs the inputs and decides whether to seek the reward. The body then acts to reach it.

Addiction

Most drugs of abuse cause huge amounts of dopamine to build up in the reward system – far more than natural rewards like food or sex. This creates a powerful drive to seek out more of the drug. It also causes the brain to reduce the number of dopamine receptors, so natural rewards no longer give the same sensation. This means the user loses the urge to to seek out things like food and social engagement. Instead, drug cues become powerful triggers for dopamine release, causing intense cravings, even when the user consciously wants to stop and no longer enjoys the drug.

UP TO **60%** OF **ADDICTION RISK** STEMS FROM **GENETIC FACTORS**

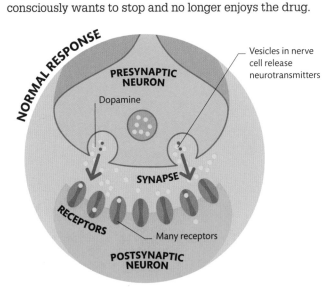

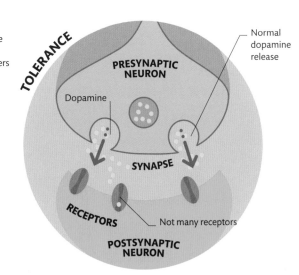

Flooded with dopamine

Some drugs of abuse increase dopamine release, while others prevent it being recycled. The build-up in the synapse produces a large response in the brain, triggering the drive to seek out more of the drug. Environmental cues become linked with the drug and can trigger cravings in the future.

Under tolerance

Over time, the brain reduces the number of dopamine receptors to counteract the excess. Now, when normal amounts of dopamine are released, they have little effect. The user may need bigger and bigger doses of the drug to feel its effect, and their desire for other rewards decreases.

WHY IS JUNK FOOD SO TASTY?

Most junk food contains lots of sugar, salt, and fat, which trigger our reward system. This would have helped us survive when food was scarce.

WANTING VERSUS LIKING

The reward pathway is often called a "pleasure pathway", and dopamine a "pleasure chemical", but this is not accurate. Dopamine in the nucleus accumbens drives "wanting" of a reward, but it is common for addicts to experience strong cravings without liking the effects of the drug. Pleasure is likely to be caused by other neurotransmitters, such as opioids or endocannabinoids.

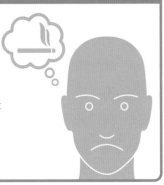

Sex and love

Sexual reproduction is fundamental to passing on our genes. Multiple emotions evolved that accompany and facilitate this process, which together can create the feeling of love.

Love and attraction

The scientific study of love and sexual behaviour has identified three primary components: attraction, attachment, and lust. These states all occur on different timescales and involve different regions of the brain producing an array of chemical messengers – neurotransmitters and hormones. Lust and attraction are closely interlinked and both are transient, passing in a relatively short time. For relationships to last, these states must yield profound attachment, which involves long-term changes to the brain.

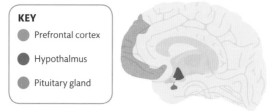

KEY
- Prefrontal cortex
- Hypothalmus
- Pituitary gland

Brain areas
The hypothalamus and pituitary gland control early hormone-led phases of bonding. The prefrontal cortex then mediates the emotional control involved in attachment.

THE LOVE DRUG

Oxytocin, which is released by the hypothalamus, has long been known as the hormone that induces labour in mammals.
It was then found to be crucial for mother-offspring bonding, and later to be central to forming long-term attachments in sexual and social relationships.

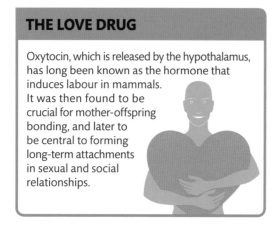

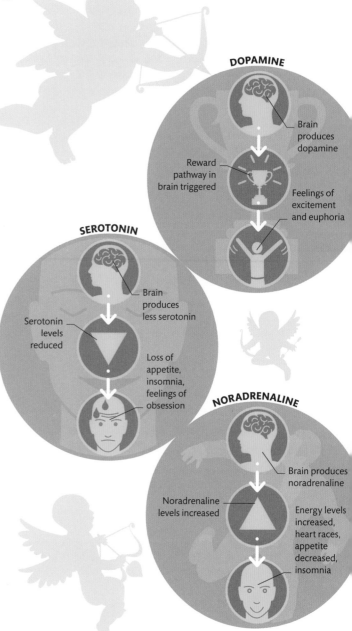

DOPAMINE
- Brain produces dopamine
- Reward pathway in brain triggered
- Feelings of excitement and euphoria

SEROTONIN
- Brain produces less serotonin
- Serotonin levels reduced
- Loss of appetite, insomnia, feelings of obsession

NORADRENALINE
- Brain produces noradrenaline
- Noradrenaline levels increased
- Energy levels increased, heart races, appetite decreased, insomnia

Attraction
Surges of the chemical messengers dopamine and noradrenaline combine with reduced levels of serotonin to produce urgent feelings of attraction. In an energized state – with racing heart, sweaty palms, and little appetite – we think constantly about our lover, craving their company.

OXYTOCIN REDUCES
ACTIVITY IN THE
BRAIN'S **FEAR** CENTRE

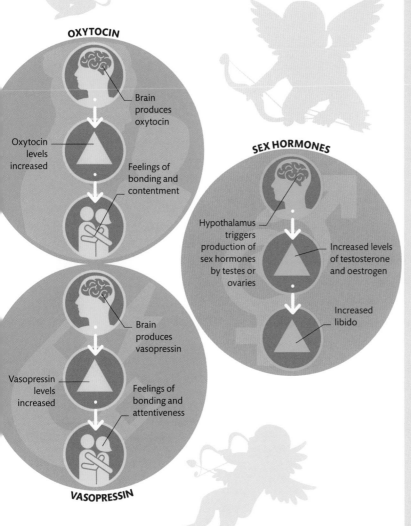

OXYTOCIN

Brain produces oxytocin

Oxytocin levels increased

Feelings of bonding and contentment

SEX HORMONES

Hypothalamus triggers production of sex hormones by testes or ovaries

Increased levels of testosterone and oestrogen

Increased libido

Brain produces vasopressin

Vasopressin levels increased

Feelings of bonding and attentiveness

VASOPRESSIN

Facial symmetry

A person's face is key to how attractive others find them. Humans and monkeys prefer symmetrical faces – symmetry is an indicator of good health and genetics. Many species also favour sexually dimorphic faces, males preferring feminine faces and vice versa. These factors interact: higher facial symmetry increases a face's perceived femininity or masculinity.

KEY
● Symmetrical face
● Asymmetrical face

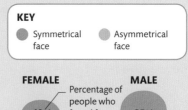

FEMALE 69% — Percentage of people who found face sex-typical — **MALE** 85%

31% 15%

European
When shown composite faces with high or low symmetry, European observers judged high-symmetry faces to appear more feminine or masculine.

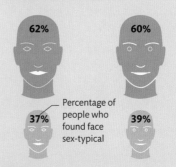

62% 60%

37% — Percentage of people who found face sex-typical — 39%

Hadza
Similar results were found in the Hadza people, an indigenous Tanzanian ethnic group. This suggests that the link between symmetry and attractiveness is universal.

Attachment
The hormones oxytocin and vasopressin have multiple effects – including making us feel more protective of our object of attraction and attentive to their needs. They stimulate long-term bond formation, but can increase distrust of others.

Lust
Lust is the primeval urge to engage in sexual relationships, driven by the sex hormones testosterone and oestrogen. While they increase libido in men and women respectively, alone they do not induce lasting connections.

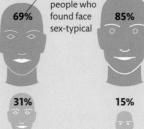

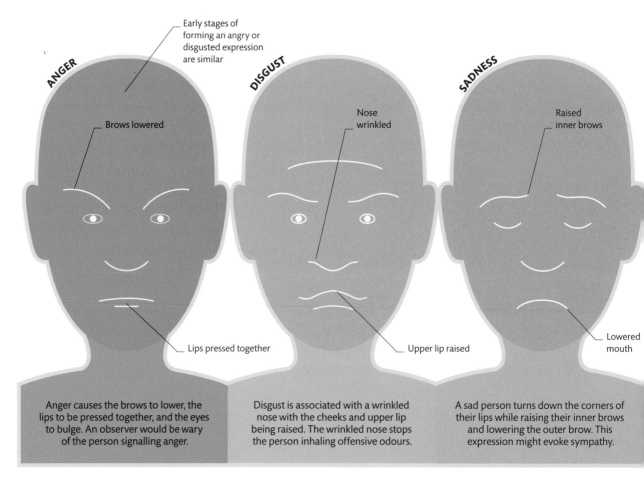

ANGER

Early stages of forming an angry or disgusted expression are similar

Brows lowered

Lips pressed together

Anger causes the brows to lower, the lips to be pressed together, and the eyes to bulge. An observer would be wary of the person signalling anger.

DISGUST

Nose wrinkled

Upper lip raised

Disgust is associated with a wrinkled nose with the cheeks and upper lip being raised. The wrinkled nose stops the person inhaling offensive odours.

SADNESS

Raised inner brows

Lowered mouth

A sad person turns down the corners of their lips while raising their inner brows and lowering the outer brow. This expression might evoke sympathy.

Universal expressions

Psychologists have found there are six universal emotions: anger, disgust, sadness, happiness, fear, and surprise. Like primary colours, they combine to give rise to the many emotions we experience. Each one is linked to a distinctive facial expression that is similar in every culture. Expressions are part biologically and part socially driven. When surprised or fearful, for example, widening the eyes takes in more light to better survey the situation. But other aspects of expressions evolved to convey social signals to members of the same species.

Expressions

Expressions are extensions of emotions. They allow us to communicate our feelings to others, and to infer the thoughts and feelings of people around us. Psychologists believe there are six basic emotions, each with an associated expression.

MICRO EXPRESSIONS

Micro expressions are tiny, involuntary, and often barely perceptible facial expressions. They last half a second or less, and the person making them may be unaware that this form of "emotional leakage" is revealing their true feelings.

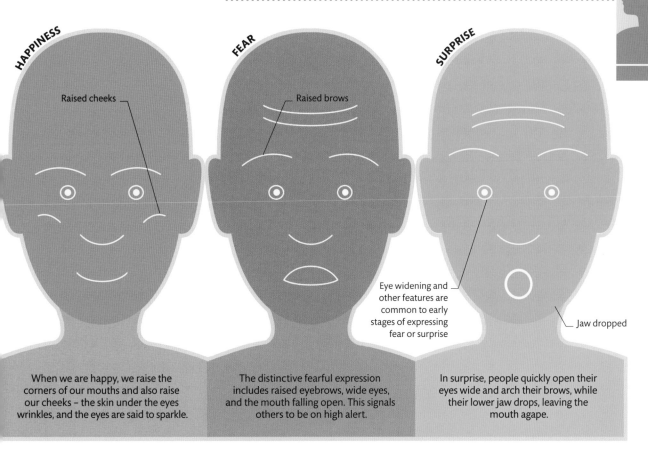

HAPPINESS

Raised cheeks

When we are happy, we raise the corners of our mouths and also raise our cheeks – the skin under the eyes wrinkles, and the eyes are said to sparkle.

FEAR

Raised brows

The distinctive fearful expression includes raised eyebrows, wide eyes, and the mouth falling open. This signals others to be on high alert.

SURPRISE

Eye widening and other features are common to early stages of expressing fear or surprise

Jaw dropped

In surprise, people quickly open their eyes wide and arch their brows, while their lower jaw drops, leaving the mouth agape.

Smiling

A smile can either be a genuine expression of positive mood or a conscious, socially motivated action. Genuine smiles are unconscious acts that involve different muscle groups to social smiles. While both involve a stretched mouth with lips upturned at the corners, the genuine smiling person constricts muscles that raise the cheeks, producing "crow's feet" around the eyes. Conscious smiles vary in their exact structure and are used in an array of social interactions – they can be socially bonding but also used to signal dominance, and people may also smile to mask embarrassment.

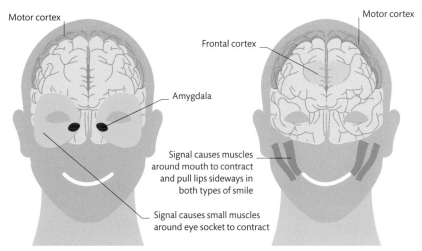

Motor cortex

Frontal cortex

Motor cortex

Amygdala

Signal causes muscles around mouth to contract and pull lips sideways in both types of smile

Signal causes small muscles around eye socket to contract

Genuine smile
The muscular contractions involved in genuine smiles are triggered by signals from the brain's emotional centres, such as the amygdala, usually operating without our awareness.

Conscious smile
Conscious control of social smiling involves activation of the frontal cortex and signals from the motor cortex. The mouth muscles contract but we can't control the eye muscles.

Body language

Body language is non-verbal communication, in which our thoughts, intentions, or feelings are expressed by physical behaviours such as body posture, gestures, eye movements, and facial expressions.

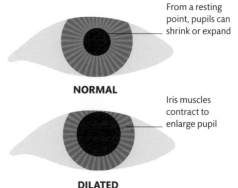

From a resting point, pupils can shrink or expand

NORMAL

Iris muscles contract to enlarge pupil

DILATED

Non-conscious communication

Social interactions between people involve complex streams of non-verbal communication that are processed in parallel to speech. Many aspects of body language arise instinctively – eye movements, facial expressions, and posture, for example, all change without conscious control. These movements can therefore reveal unspoken intentions. Body language is also used to signal social intentions overtly, such as when blowing a kiss. The richness of this communication involves the whole body and our brains are attuned to it.

Eye signals
Pupils frequently shift size and can signal various things. A dilated pupil may indicate surprise or attraction. Constricted pupils are associated with negative emotions such as anger.

HAPPY

AGGRESSIVE

MORE THAN **50 PER CENT** OF COMMUNICATION IS BASED ON OUR **BODY LANGUAGE**

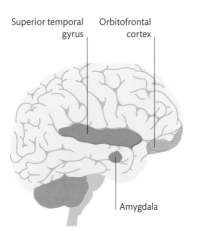

Superior temporal gyrus

Orbitofrontal cortex

Amygdala

Brain processes
Processing body language involves areas like the amygdala, which receives emotional content; part of the superior temporal gyrus, which responds to seeing human movement; and the orbitofrontal cortex, which analyses meaning. Special cells, called mirror neurons (see pp.102–03), are also activated when you see someone else moving.

DO GESTURES HAVE THE SAME MEANING AROUND THE WORLD?

No, many gestures are culturally specific. A simple hand gesture can have different meanings for different societies.

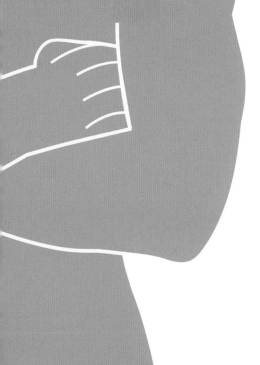

SAD

DEFENSIVE

Facial expressions

Facial expressions reveal much about a person's emotions (see pp.116–17). The eyes and the mouth, in particular, automatically respond to strong feelings, although people can consciously change their expressions to mask emotions.

Posture

An aggressive posture tends to inflate a person's size. It may involve extending the arms, setting the feet far apart, and protruding the chest. The same postures may be used to invade others' personal space. In contrast, defensive postures are closed – folded arms, for example, are a classic indicator.

Gestures

Most body language is performed unconsciously, but we have more conscious control over our gestures, which are movements of the body used to convey meaning. There are four categories of gestures: symbolic (or emblematic); deictic (or indexical); motor (or beat); and lexical (or iconic). They might be used instead of speech or alongside it for emphasis. Some scientists believe that increasingly complex gestures evolved as the forerunners of speech, which now defines our species.

TYPES OF GESTURE

Symbolic
These are gestures that can be literally translated into words – for example, waving hello or making the "ok" sign. They are widely recognized in a given culture but may not be recognized beyond that culture.

Deictic
Deictic gestures involve pointing or otherwise indicating a concrete object, person, or more intangible item. Used with or without speech, they act like pronouns, meaning "this" or "that".

Motor
This type of gesture is short and tied to speech patterns, such as moving the hand in time with speech, and is used for emphasis. Motor gestures contain no inherent meaning and are meaningless without accompanying vocalization.

Lexical
These gestures depict actions, people, or objects, such as miming throwing when telling a story about throwing a ball, or using your hands to depict an object's size. They usually accompany speech but contain meaning independently.

SIGN LANGUAGE

Sign language may appear to be a sophisticated type of body language, but it has more in common with speech. Studies show that when people sign, the same brain areas (see right) light up as when they speak. Sign language has grammar, and each gesture has a specific meaning, while body language is interpreted broadly.

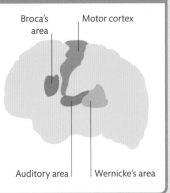

Broca's area

Motor cortex

Auditory area

Wernicke's area

How to tell if someone is lying

Separating truth from falsehood depends partly on knowing a person, so you can judge if they are behaving differently from usual. With a confident and persuasive talker, especially someone you don't know, how easy is it to spot a lie?

The short answer is: it is difficult. Traditional telltale signs of lying are shifting gaze to avoid eye contact, folding and unfolding arms, shrugging shoulders, and fidgety hands and feet. However, scientific studies do not support these beliefs. Some honest people are generally nervous and squirmy. In others, these signs show someone is concentrating on being trustworthy.

Polygraph, or "lie detector", machines – which record pulse and breathing rates, blood pressure, and sweating – have a dubious history. This is partly due to the stress of using them. Innocent but anxious people can show up as deceitful, while calm, skilled liars pass easily.

Clues from speech

Speech can be slightly more reliable. Hesitation, repeated words or phrases, breaking up sentences, a change in tone or in speaking speed, vagueness, and describing trivial details while avoiding the main topic – are all strategies to give the brain "time to think" and work out which falsehood might be most believable. This is especially true for persistent liars, who must access memory so as not to contradict themselves as their multiple deceptions become ever more tangled.

A more reliable method involves the use of fMRI (see p.43), a brain scan that requires the person's total cooperation. Certain parts of the brain are more active when lying and show up together on screen. These include the prefrontal, parietal, and anterior cingulate cortices and the caudate nucleus, thalamus, and amygdala.

In summary:
- **Be very aware of judging someone you don't know well.**
- **Don't rely on time-honoured signs such as fidgeting and lack of eye contact.**
- **Clues from speech, such as hesitation and repetition, can be slightly more reliable.**
- **In many tests, a simple "gut feeling" was as successful as most other methods.**

Morality

Most people living in normal environments develop instinctive senses of right and wrong. Morality seems to be in part hard-wired, arising from the conjunction of rationality and emotion.

Where do right and wrong come from?

Social norms based on shared morals exist across all cultures, enabling social cohesion. When making moral decisions, two brain systems come into play: a "rational" system that effortfully and explicitly weighs the pros and cons of possible actions; and a system that rapidly generates emotional, intuitive feelings of right and wrong. Interactions between rationality and emotion are complex, but studying brain activity while people grapple with moral dilemmas has identified the key areas involved.

Moral judgement
When we make decisions, our emotions play a vital role. In order to weigh up moral matters, brain areas that are involved in emotional experience coordinate with areas that register facts and consider possible actions and consequences.

KEY

 Rational circuit

Emotional circuit

Parietal lobe
Involved in working memory and cognitive control, this area of the cortex provides information needed to help us perceive social signals, to work out others' beliefs and intentions – such as whether an act was aggressive or how a social context should affect behaviour.

Dorsolateral prefrontal cortex
This area integrates rational and emotional information. It may also counteract the ventromedial area to suppress emotional drives when dealing with complex moral dilemmas that favour cognitive solutions using memories or other data.

Amygdala

Posterior superior temporal sulcus
This part of the cortex functions with the parietal lobe, providing information to guide moral intuition and attributing beliefs to others, and integrating this data with the potential outcomes of actions. It also helps assess if a person is lying.

EXTERNAL VIEW

Ventromedial prefrontal cortex
This area is an important structure for allowing emotional responses to influence rationalized moral decisions. In psychopaths, connections between this region and both the amygdala and reward pathways are disrupted.

Temporal pole
The temporal pole functions in both social processing, such as face recognition and working out the mental states of others, and in emotional processing. It may also help to combine complex perceptual inputs with intuitive emotional responses.

Altruism

Altruism – when a person acts to benefit another at personal cost or risk – involves empathizing with another's distress then acting to help. It involves distinct processes. Brain scans show that acting altruistically activates the reward pathways (see pp.112–13), reinforcing the behaviour and quelling emotional discomfort. Selflessness is a distinguishing feature of human behaviour and an evolutionary enigma given dangers to the altruist.

PSYCHOPATHY

Psychopaths can understand morality and can, therefore, mimic normal social interactions. This means that while they behave heinously, they remain hard to identify. The underlying cause may be a disconnect between brain regions linking logical decision-making and emotion, leaving them unable to grasp the fallout from their behaviour.

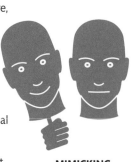

MIMICKING EMOTIONS

Posterior cingulate cortex

This region is active when our environment changes and when we are thinking about ourselves. It may help assess the seriousness of offences and the appropriate response by acting as a hub for integrating intuitions about the mental states of others.

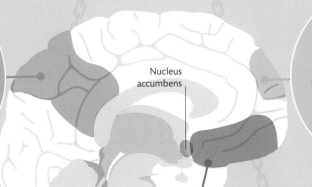

Nucleus accumbens

Medial frontal gyrus

This region of the brain is important for decision-making and for choosing between alternative potential actions. This is especially the case when there is conflict between multiple options.

INTERNAL VIEW

SEEING SOMEONE **HURT BY ACCIDENT** PRODUCES SIMILAR **BRAIN ACTIVITY** AS IF THE **VIEWER WAS HURT THEMSELVES**

Orbitofrontal prefrontal cortex

Activated by watching morally charged scenes, this area processes emotional stimuli. It aids in representing just rewards and punishments for observed behaviour and in making emotionally driven moral choices.

CAN BRAIN DAMAGE AFFECT MORALITY?

It depends on the area affected. For example, damage to regions that link emotion to moral choice can cause people to make "cold-hearted" decisions.

Learning a language

Unlike other species, humans have a brain with regions dedicated to language. Babies are born ready to learn language, acquiring it through an interplay between these specialized areas of the brain and their own unique experiences. To learn language, we also have to interact with other people.

Learning to talk

Our innate preference for looking at faces helps newborns to focus attention on people talking to them. Later, making eye contact and following gaze allows them to connect the words they hear with what is being talked about. As they learn new words, infants make "overextension" errors by using a single word to label multiple things, for example, by using the word "fly" to refer to anything small and dark.

Timeline of speech

The exact timescale for mastering language varies from child to child, but all children progress through the main stages in a similar order – from cooing and babbling to first words and, ultimately, full sentences.

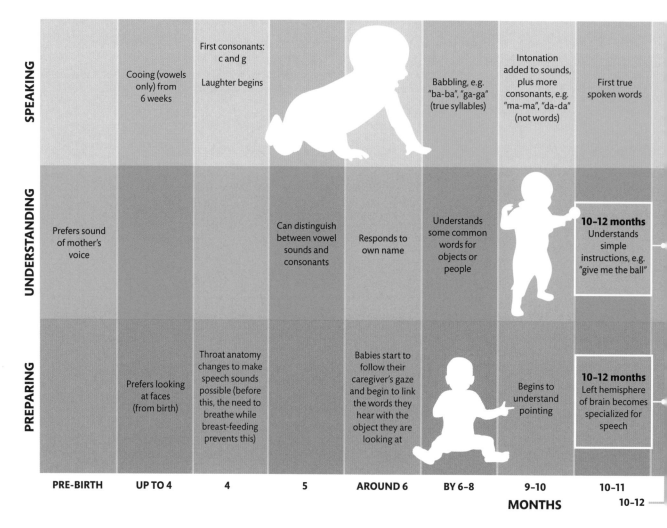

	PRE-BIRTH	UP TO 4	4	5	AROUND 6	BY 6–8	9–10	10–11 / 10–12
SPEAKING		Cooing (vowels only) from 6 weeks	First consonants: c and g; Laughter begins			Babbling, e.g. "ba-ba", "ga-ga" (true syllables)	Intonation added to sounds, plus more consonants, e.g. "ma-ma", "da-da" (not words)	First true spoken words
UNDERSTANDING	Prefers sound of mother's voice			Can distinguish between vowel sounds and consonants	Responds to own name	Understands some common words for objects or people		**10–12 months** Understands simple instructions, e.g. "give me the ball"
PREPARING		Prefers looking at faces (from birth)	Throat anatomy changes to make speech sounds possible (before this, the need to breathe while breast-feeding prevents this)		Babies start to follow their caregiver's gaze and begin to link the words they hear with the object they are looking at		Begins to understand pointing	**10–12 months** Left hemisphere of brain becomes specialized for speech

MONTHS

The bilingual brain

In the brain of a bilingual speaker, languages "compete" for attention. This provides unconscious practice in ignoring irrelevant information, and studies show that bilinguals are better at this than monolinguals. The ability to learn a second language like a native speaker is usually lost after around four years of age, especially with pronunciation. The brains of elderly bilinguals show better preservation of white matter, which may protect them from the effects of cognitive decline.

White matter preserved in older bilingual adults

RIGHT HEMISPHERE

Activated region of grey matter

LEFT HEMISPHERE

Bilingualism areas
Areas of grey matter (shown in blue) are activated in bilingual speakers when they switch between languages.

ALCOHOL AND LANGUAGE

One study of second-language learners looked at whether alcoholic drinks would improve speaking and pronunciation by reducing self-consciousness. It worked up to a point – but after too many drinks, performance rapidly deteriorated.

BONJOUR, ÇA VA?

BHLEES CHIDEVSSSS

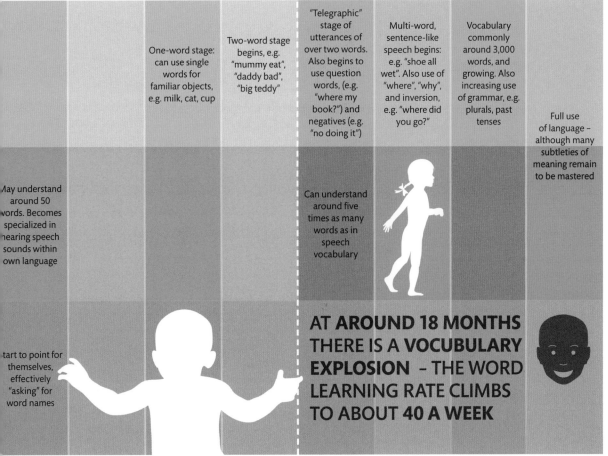

One-word stage: can use single words for familiar objects, e.g. milk, cat, cup

Two-word stage begins, e.g. "mummy eat", "daddy bad", "big teddy"

"Telegraphic" stage of utterances of over two words. Also begins to use question words, (e.g. "where my book?") and negatives (e.g. "no doing it")

Multi-word, sentence-like speech begins: e.g. "shoe all wet". Also use of "where", "why", and inversion, e.g. "where did you go?"

Vocabulary commonly around 3,000 words, and growing. Also increasing use of grammar, e.g. plurals, past tenses

Full use of language – although many subtleties of meaning remain to be mastered

May understand around 50 words. Becomes specialized in hearing speech sounds within own language

Can understand around five times as many words as in speech vocabulary

Start to point for themselves, effectively "asking" for word names

AT **AROUND 18 MONTHS** THERE IS A **VOCABULARY EXPLOSION** – THE WORD LEARNING RATE CLIMBS TO ABOUT **40 A WEEK**

| AROUND 12 | FROM 12 | 12–18 | 18 MONTHS | 2 YEARS | 2–2.5 | 3 ONWARDS | 5 |

YEARS

The language areas

The human brain, unlike that of any other animal, has areas dedicated specifically to language, usually located in its left hemisphere. The unique ability of humans to communicate using language is thought to be an evolutionary advantage.

Broca's and Wernicke's areas

The two main language areas are Broca's and Wernicke's areas. Broca's area is associated with moving the mouth to articulate words. When learning new languages, separate parts of Broca's area are activated when we speak either our native or non-native tongue. In Wernicke's area, words that we hear or read are understood and selected for articulation as speech. Damage to this part of the brain can lead people to speak in peculiar ways, creating sentences that do not make sense.

Motor cortex
The motor cortex enables the physical movements required to produce language – for example, moving your tongue, lips, and jaw. The motor cortex is activated when words that are semantically related to body parts are heard or spoken. For example, the word "dance" might be related to your feet.

Speech travels through air as sound waves

BRAIN DAMAGE AND LANGUAGE CHANGES

There have been cases in which patients with brain injury appeared to wake up speaking a different language or with a different accent. Foreign accent syndrome is one example of such a medical condition. These cases are rare and there have not been sufficient scientific studies carried out to understand them in any detail.

`# & @ å ž ø ï ¿ œ » § ë`

HELLO

SHWMAE   BONJOUR

ASALAAM ALAIKUM

GUTEN TAG

PRIVET   OLÁ

KONNICHIWA

HOLA   CIAO

Speaking and understanding
Processing language is a complex task. Articulating or decoding even a simple greeting, such as "hello" requires several different areas of the brain to work together.

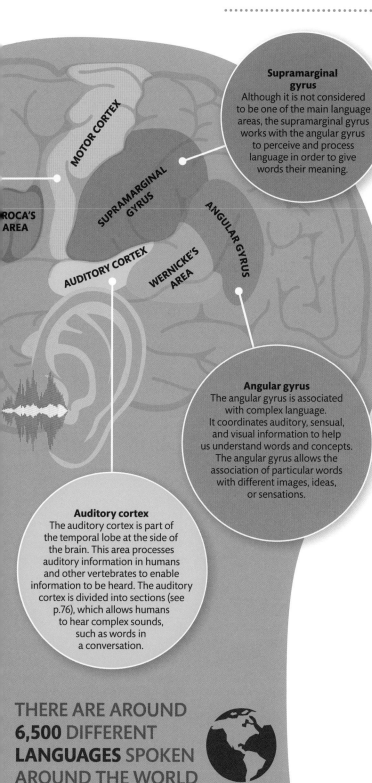

Supramarginal gyrus
Although it is not considered to be one of the main language areas, the supramarginal gyrus works with the angular gyrus to perceive and process language in order to give words their meaning.

MOTOR CORTEX

BROCA'S AREA

SUPRAMARGINAL GYRUS

ANGULAR GYRUS

AUDITORY CORTEX

WERNICKE'S AREA

Angular gyrus
The angular gyrus is associated with complex language. It coordinates auditory, sensual, and visual information to help us understand words and concepts. The angular gyrus allows the association of particular words with different images, ideas, or sensations.

Auditory cortex
The auditory cortex is part of the temporal lobe at the side of the brain. This area processes auditory information in humans and other vertebrates to enable information to be heard. The auditory cortex is divided into sections (see p.76), which allows humans to hear complex sounds, such as words in a conversation.

THERE ARE AROUND 6,500 DIFFERENT LANGUAGES SPOKEN AROUND THE WORLD

Aphasia

Aphasia is a medical condition in which people are unable to comprehend or produce language, read, or write due to damage caused to the brain – for example as the result of a trauma, stroke, or tumour. The condition can be relatively mild or severe. There are many types of aphasia (for some examples see table, below). Some are named after the brain area that is affected or the type of speech produced. However, aphasia can affect language, reading, and writing in many different ways, and some of these difficulties may not fit into one specific type or category.

TYPES OF APHASIA	
TYPE	**SYMPTOMS**
Global	The most severe form of aphasia, causing general deficits in comprehension, understanding, and production of language.
Broca's	Speech production is affected and can be reduced to just a few words, which may be halting or "non-fluent" in their nature.
Wernicke's	An inability to understand the meaning of words. Speech production is unaffected but irrelevant words may be used, forming nonsensical phrases.
Anomic	Difficulty finding words during speaking or writing. This can lead to vague language, causing significant frustration.
Primary progressive	Language capabilities become slowly, progressively impaired. This form can be caused by diseases such as dementia.
Conduction	A rare form of aphasia that causes difficulty repeating phrases, particularly if phrases or sentences are long and complex.

Facial expressions

We constantly use facial expressions during conversation. As speakers, we raise eyebrows to emphasize a point or indicate a question, and as listeners we use expressions to show interest in what is being said. One study looked at the top reasons for using facial expressions in conversation.

 FACIAL SHRUG

 THINKING

 **EMPHASIS**

 **EMPATHIC**

 **QUESTION**

 **RETELLING**

 **PERSONAL REACTION**

I'M LISTENING

KEY
- Speaker
- Listener
- Both

THE SPEAKER

1 Message idea
The starting point of a conversation is an idea the speaker wants to express and the intention to express it.

2 Formulation
The speaker selects the words with the right meaning (semantics) and then puts them into the right form and order (syntax) to make sense. For example, "Would you like a drink?" is a question, "You would like a drink" is a statement, and "Like you drink a would" is nonsense. Broca's area (see p.126) is crucial to these two processes.

LIKE WOULD YOU
SEMANTICS

WOULD YOU LIKE
SYNTAX

3 Articulation
To say the message, the speaker moves the mouth, tongue, lips, and throat, controlled by the motor cortex, to form the speech sounds with the right intonation.

NO, THANKS
TURN TAKING

WOULD YOU LIKE A DRINK?

GARDEN PATH SENTENCES

We can be misled if the first part of a message suggests an idea that is contradicted by the later part. For example: "The car stopped at the crash scene was soon surrounded by police." We initially understand "stopped" to mean something the car did; but when we hear "was soon", it becomes clear that the car was stopped by police. We have to revisit the start of the message to make sense of it. This type of statement is called a garden path sentence.

Having a conversation

A conversation is a shared endeavour between speaker and listener, which involves more than producing and understanding words. We take turns, signal understanding, and align our thoughts.

Beyond words

We constantly use non-verbal signals alongside speech in conversation. As well as adding emphasis (via facial expressions) or visual effect (via gestures), such signals allow the person not speaking to have a role in the conversation partnership, encouraging the speaker without interrupting or taking over.

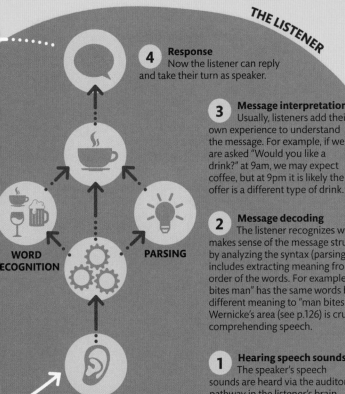

THE LISTENER

4 Response
Now the listener can reply and take their turn as speaker.

3 Message interpretation
Usually, listeners add their own experience to understand the message. For example, if we are asked "Would you like a drink?" at 9am, we may expect coffee, but at 9pm it is likely the offer is a different type of drink.

2 Message decoding
The listener recognizes words and makes sense of the message structure by analyzing the syntax (parsing). Parsing includes extracting meaning from the order of the words. For example, "dog bites man" has the same words but different meaning to "man bites dog". Wernicke's area (see p.126) is crucial in comprehending speech.

1 Hearing speech sounds
The speaker's speech sounds are heard via the auditory pathway in the listener's brain.

WORD RECOGNITION   PARSING

Speaking and listening

Speaker and listener swap roles many times in a conversation – and as speakers, we also monitor our own speech output. Although both roles involve multiple steps, it can all happen fast – taking from 0.25 seconds between having an idea to saying it, and from 0.5 seconds for comprehension. Hesitation occurs when speakers need time to "catch up" with the complex speech planning and production process.

ELEMENTS OF CONVERSATION

Looking
Listeners look at their conversation partner much more than speakers do. They do this to show interest – as without this, speakers often falter. In contrast, speakers look intermittently at the listener.

Gestures
We use many types of hand gestures (see p.119), including: conventional signs – such as "thumbs up"; pointing; and expressive hand movements, to add emphasis to the message.

"I'm listening" signals
Listeners use non-verbal sounds and gestures, such as saying "mmm" or nodding, to show they are engaged in the conversation while not speaking.

Turn-taking
Conversation requires taking turns, and we start learning this from infancy. Conversation partners rarely talk over each other, even though the average gap between turns is only a few tenths of a second.

PEOPLE TALK OVER EACH OTHER LESS THAN 5% OF CONSERVATION TIME

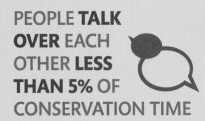

Learning to read and write

The ability to read and write is something that most people start to learn at a young age. As our brains develop, we learn important reading and writing skills. By the time we reach adulthood, we can read on average 200 words per minute. Reading requires several areas of the brain and body to work together. For example, when you read, your eyes need to recognize the word on a page and your brain then processes what that word says. Writing uses the brain's language areas (see pp.126–27), visual areas, and motor areas concerned with manual dexterity to make the necessary hand movements.

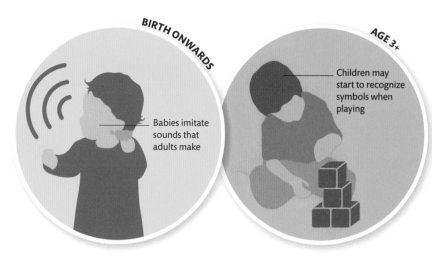

BIRTH ONWARDS

Babies imitate sounds that adults make

AGE 3+

Children may start to recognize symbols when playing

1 Making sounds
Babies make sounds that imitate adults but often aren't recognizable as words. This is the foundation for learning to develop language skills. Babies see and process facial expressions using the visual cortex and other areas. They then learn to associate sounds and facial expressions with things in the world.

2 Recognizing symbols
Children begin to understand what symbols mean when they are in text. They use the visual cortex and memory to translate symbols that they see into sounds. As children grow, they connect these sounds with the meanings of words and start to relate language to written text.

Reading and writing

Our brains are hardwired for speech but the ability to read and write is not innate. We have to start training our brains as babies to develop these complex skills.

WHAT CAUSES DYSLEXIA?

Research suggests children with dyslexia have problems understanding the sounds letters make, but dyslexia is also found in cultures where symbols represent an idea rather than a sound.

DYSGRAPHIA

Dysgraphia is the inability to write clearly. It can be the symptom of some brain conditions, such as Parkinson's disease, that affect fine motor skills. Writing may be wobbly and indistinct or completely mangled.

tHisIsaS eNT EncEwriT TtENbY sOMEonEwItHdYsGRapHiA

SPEED READERS ARE ABLE TO READ MORE THAN **700** WORDS PER MINUTE

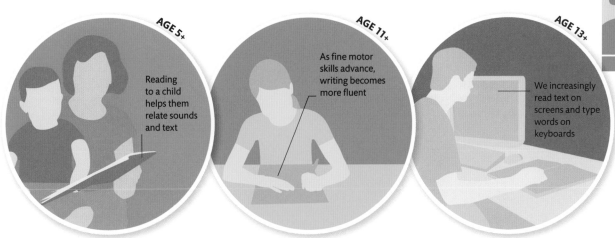

AGE 5+

Reading to a child helps them relate sounds and text

AGE 11+

As fine motor skills advance, writing becomes more fluent

AGE 13+

We increasingly read text on screens and type words on keyboards

3 Beginning to read
Reading aloud can improve a child's reading ability. Listening to a story activates the auditory cortex to hear the words, which are then processed by the frontal lobe. Picture books help children practise relating words to images, and asking them to join in reading builds vocabulary and comprehension.

4 Expanding vocabulary
As we grow older, we experience more of the world around us so we learn and see new things, adding to our vocabulary. Comprehension, the ability to understand how to use words, requires every lobe of the brain (see p.30) and the cerebellum to successfully comprehend and use language.

5 Continuing to learn
As adults, we continue to learn and practise our reading and writing skills. Our vocabulary is constantly being extended. Learning to read and write is just the start of the story. The whole brain is required to maintain language skills, and good brain health is vital to both reading and writing.

Dyslexia

Dyslexia takes various forms, affecting people's ability to read or write, or both. It is thought that up to one in five people have dyslexia. A full neurological explanation of the causes of dyslexia has not yet been achieved. Studies have suggested that particular structures of the brain function differently in dyslexia (see right). As children with dyslexia typically struggle with their reading abilities, it is difficult to determine if the developing brain impacts the dyslexia or if the dyslexia itself has an impact on the developing brain.

Non-dyslexic brain reading
Broca's area helps form and articulate speech. The parietal-temporal cortex works to analyse and understand new words. The occipital-temporal area forms words and aids in meaning, spelling, and pronunciation.

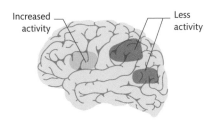

Increased activity

Less activity

Dyslexic brain reading
Broca's area is activated to form and articulate words but the parietal-temporal and occipital–temporal areas are less active. Broca's area can be over-activated to compensate for the lack of stimulation of the other regions.

KEY

● Parietal-temporal
● Occipital-temporal
● Inferior frontal gyrus (Broca's area)

ALPHABETIC PRINCIPLE

The alphabetic principle is the idea that individual letters or groups of letters represent sounds when they are spoken aloud. The alphabetic principle has two parts:

1. Alphabetic understanding
Learning that words are made up of letters that represent the sounds made when speaking these letters aloud.

2. Phonological recoding
Understanding how strings of letters in written words combine to make sounds, which enables spelling and pronunciation.

MEMORY, LEARNING, AND THINKING

What is memory?

Our memory allows us to learn from experience and shapes us as individuals. Memory is not a single discrete brain function, there are several types, involving different brain areas and processes.

Memory in the brain

Memory includes instinctive processes that you are unaware of, as well as the more obvious parts that allow you to remember what you had for lunch yesterday or your boss's name. Each type of memory uses a range of different brain areas. Scientists used to think the hippocampus was vital for all new memories to form, but now it is thought this is only the case for episodic memories. Other types of memory use other areas, which are spread all around the brain.

Types of memory

To better understand how it works, scientists break memory down into a number of types. Many of these rely on different networks within the brain, although there is also a lot of overlap between the brain areas involved in each category.

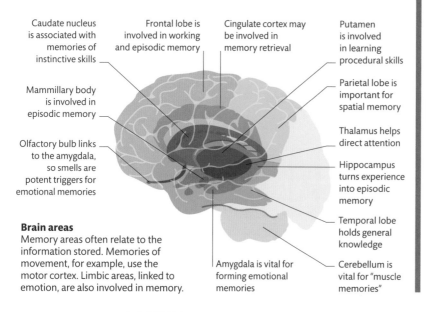

Caudate nucleus is associated with memories of instinctive skills

Frontal lobe is involved in working and episodic memory

Cingulate cortex may be involved in memory retrieval

Putamen is involved in learning procedural skills

Mammillary body is involved in episodic memory

Parietal lobe is important for spatial memory

Thalamus helps direct attention

Olfactory bulb links to the amygdala, so smells are potent triggers for emotional memories

Hippocampus turns experience into episodic memory

Brain areas
Memory areas often relate to the information stored. Memories of movement, for example, use the motor cortex. Limbic areas, linked to emotion, are also involved in memory.

Temporal lobe holds general knowledge

Amygdala is vital for forming emotional memories

Cerebellum is vital for "muscle memories"

Short-term memory
Short-term memory is very limited – only storing around 5–9 items, but this varies between individuals and for different types of information. To keep something in short-term memory, we often repeat it to ourselves, but if we are distracted, we instantly forget it.

Non-associative learning
When you are repeatedly exposed to the same stimulus, such as a light, a sound, or a sensation, your response changes. For example, when you come home, you smell dinner cooking, but gradually the smell seems to fade. This is known as habituation, one form of non-associative learning.

Simple classical conditioning
Made famous by the Russian physiologist Ivan Pavlov and his dogs, in classical conditioning repetition causes something neutral to be linked with a response. An example is your mouth watering as you enter a cinema lobby, as you have learned to expect popcorn in that environment.

Priming and perceptual learning
In priming experiments, you are shown a word or picture so quickly you don't consciously "see" it – but it can still affect your behaviour. For example, someone primed with the word "dog" will recognize the word "cat" faster than a completely unrelated word such as "tap".

Memory systems
Memory is split into two main types: short- and long-term memory. Short-term memories are fleeting, but important information can be passed over to long-term memory for storage. Long-term memories may last a whole lifetime and are further divided into several different types of memory.

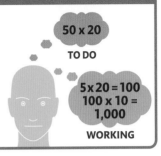

WORKING MEMORY
To multiply 50 x 20 you must manipulate the numbers stored in short-term memory. This uses a process called working memory. Working memory ability is one of the best predictors of success in school for young children.

50 x 20
TO DO

5 x 20 = 100
100 x 10 = 1,000
WORKING

Long-term memory
Our long-term memory allows us to store a – theoretically – almost infinite number of memories for most of our life. Long-term memories are stored as distributed networks of neurons spread out across the outer layer of the brain, the cortex. Recalling the memory sparks the network to fire again.

Non-declarative (implicit)
Non-declarative memories are unconscious, so cannot be passed from person to person using words. You might try, for example, to explain to someone how to tie their shoe laces or ride a bike, but they would probably still fail or fall off the first time they attempted to do it for themselves.

Declarative (explicit)
Declarative memories can be told to someone else. They are conscious and sometimes learned through repetition and effort, although others can be stored without awareness of the process. They include memories of events that have happened in your life (episodic) and facts (semantic).

Procedural
Skills or abilities, such as riding a bike or dancing, are termed procedural memories. When first learned, they require concentration and conscious effort but over time they become habit. Often called "muscle memory", procedural memories are actually stored in a brain network involving the cerebellum.

Episodic
Episodic memories might be recalling a big event like your 18th birthday or something mundane like yesterday's breakfast. These are things you actually remember happening: recalling an episodic memory is almost like reliving the event. The hippocampus is vital for storing new episodic memories.

Semantic
Semantic memories are factual – meaning they are things that you know rather than things that you remember. For example, these might include recalling the capital of France or the first three digits of Pi. Semantic memory relies on a large network of brain areas and may not involve the hippocampus at all.

How a memory forms

When networks of neurons in the brain are repeatedly activated, changes in the cells strengthen their connections, making it easier for each to activate the next (see pp.26–27). This process is known as long-term potentiation.

Strengthening connections

When you repeatedly activate a group of neurons – by practising a skill or revising facts, for example – they begin to change. This is how we form long-term memories (see p.135) in a process called long-term potentiation, which depends on various mechanisms taking place in brain cells. The first (presynaptic) neuron makes more neurotransmitter to release when the signal reaches it, and the second inserts more receptors into its membrane. This speeds up transmission at the synapse. Something like driving a car, which seems complex when you start, can become effortless as the neural pathways involved become more efficient. If this paired activation is repeated enough, new dendrites can grow, linking the two neurons via new synapses, giving the message alternative pathways and helping it travel even faster.

MEMORY TRACES

Scientists have recently been able to pinpoint a precise memory trace in someone's brain. In general, memories tend to be stored near the area of the brain that relates to how they were formed. For example, memories for voices would be near the language centres, and things that you have seen are stored, at least partly, near the visual cortex.

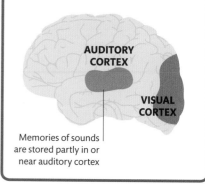

AUDITORY CORTEX

VISUAL CORTEX

Memories of sounds are stored partly in or near auditory cortex

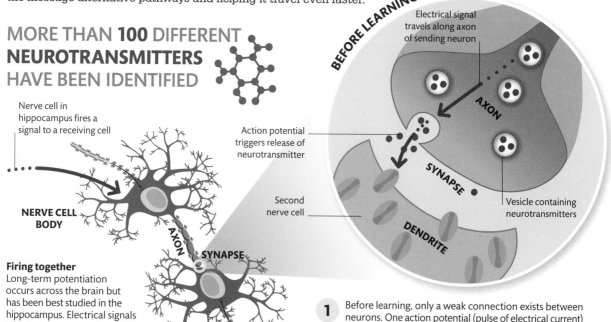

BEFORE LEARNING

Electrical signal travels along axon of sending neuron

AXON

Action potential triggers release of neurotransmitter

SYNAPSE

Second nerve cell

Vesicle containing neurotransmitters

DENDRITE

Nerve cell in hippocampus fires a signal to a receiving cell

NERVE CELL BODY

AXON

SYNAPSE

Firing together

Long-term potentiation occurs across the brain but has been best studied in the hippocampus. Electrical signals travel along a neuron's axon to the synapse, where chemical messengers are released.

1 Before learning, only a weak connection exists between neurons. One action potential (pulse of electrical current) from the first cell releases only a small amount of neurotransmitter, and this may or may not be enough to activate the next neuron, which has just a few receptors.

Emotional memories

When something strongly emotional happens, whether that is good or bad, stress chemicals such as adrenaline and noradrenaline are released. These make it easier for long-term potentiation to occur with fewer repetitions. This explains why emotionally arousing memories are stored more rapidly in the brain, and why they are easier to recall than non-emotional memories.

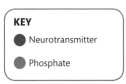

KEY
- Neurotransmitter
- Phosphate

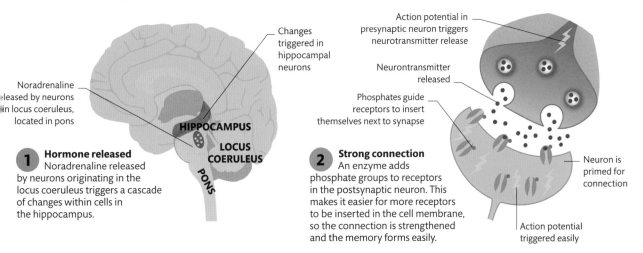

Changes triggered in hippocampal neurons

Noradrenaline released by neurons in locus coeruleus, located in pons

HIPPOCAMPUS
LOCUS COERULEUS
PONS

Action potential in presynaptic neuron triggers neurotransmitter release

Neurontransmitter released

Phosphates guide receptors to insert themselves next to synapse

Neuron is primed for connection

Action potential triggered easily

1 Hormone released
Noradrenaline released by neurons originating in the locus coeruleus triggers a cascade of changes within cells in the hippocampus.

2 Strong connection
An enzyme adds phosphate groups to receptors in the postsynaptic neuron. This makes it easier for more receptors to be inserted in the cell membrane, so the connection is strengthened and the memory forms easily.

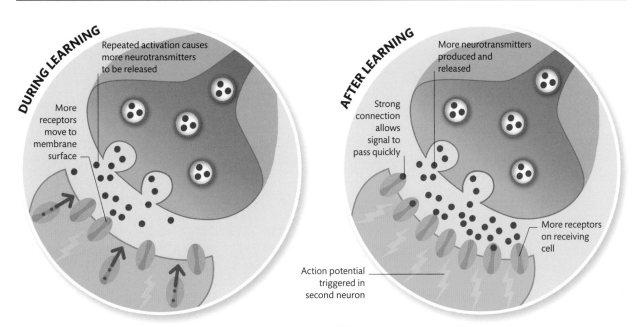

DURING LEARNING

Repeated activation causes more neurotransmitters to be released

More receptors move to membrane surface

AFTER LEARNING

More neurotransmitters produced and released

Strong connection allows signal to pass quickly

More receptors on receiving cell

Action potential triggered in second neuron

2 Both neurons firing repeatedly at the same time causes a chemical cascade within the second cell (see p.26), which makes it more sensitive to the neurotransmitter, and causes extra receptors to migrate to the edge of the synapse. A signal travels back to the first cell, telling it to produce more neuotransmitter.

3 Now, a single action potential causes the release of more neurotransmitter, carrying the message quickly and efficiently across the synapse, where it is received by many receptors. This makes it easier for the second neuron to be activated, sending its electrical signal onwards.

Storing memories

After being encoded by the hippocampus, memories are consolidated and transferred to the cortex for long-term storage. These memories are formed by strengthening connections, a process called long-term potentiation (see pp.136–137).

(see pp.136–137)

Storage in the cortex

To transfer memories for long-term storage, the hippocampus repeatedly activates a network of connections in the cortex. Each activation strengthens connections until they are secure enough to store the memory. It was thought that memories formed first in the hippocampus, with the cortical memory trace forming later, but recent research in mice suggests that they may form simultaneously, although the cortical memory is initially unstable. Repeated reactivation of the network somehow "matures" the cortical memory, meaning we can use it.

WHY DO I FORGET WHERE I LEFT MY KEYS?

Often, things we "forget" actually weren't stored as memories in the first place, because we weren't paying attention when we did them.

CORTEX

PREFRONTAL CORTEX

Memory bank
Memories are stored as networks of connections in the cortex. The number of neurons here creates a near infinite amount of possible combinations – in theory, long-term memory is virtually unlimited.

Consolidation

This storage process, known as consolidation, happens mainly while we sleep. During this time, your brain is not processing information from the outside world, so it can carry out these housekeeping tasks. Memories are sorted, prioritized, and the gist extracted. They are also linked with older memories, already in storage. This makes it easier to retrieve important memories in the future. Studies have shown it is better to take a nap after learning something new than it is to keep studying!

LEARNING

1 Study
When you learn something new, your brain takes in that information and forms new connections, or strengthens synapses that already exist.

CONSOLIDATION

2 Sleep
While you sleep, new information is consolidated. The memory becomes less reliant on the hippocampus, and less likely to be affected by interference from other inputs or brain injury

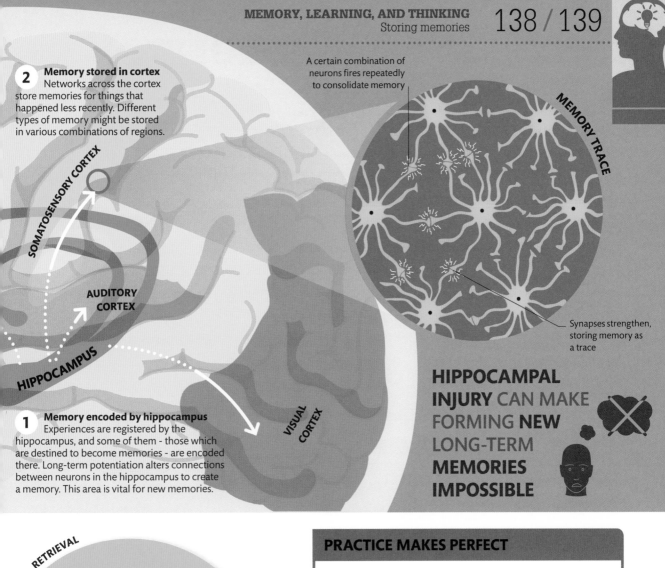

2 **Memory stored in cortex**
Networks across the cortex store memories for things that happened less recently. Different types of memory might be stored in various combinations of regions.

SOMATOSENSORY CORTEX

AUDITORY CORTEX

HIPPOCAMPUS

VISUAL CORTEX

1 **Memory encoded by hippocampus**
Experiences are registered by the hippocampus, and some of them - those which are destined to become memories - are encoded there. Long-term potentiation alters connections between neurons in the hippocampus to create a memory. This area is vital for new memories.

A certain combination of neurons fires repeatedly to consolidate memory

MEMORY TRACE

Synapses strengthen, storing memory as a trace

HIPPOCAMPAL INJURY CAN MAKE FORMING NEW LONG-TERM MEMORIES IMPOSSIBLE

RETRIEVAL

3 **Remember**
When you wake up, the memory of what you learned is stored more securely. It has also been linked to other facts, making it easier to recall, and you may find that you understand the underlying concepts better.

PRACTICE MAKES PERFECT

If you learn something just once, over time that memory trace will fade as the connections weaken. The more times you practice or revise something, the stronger those connections between neurons become and the more likely you are to remember it in the future.

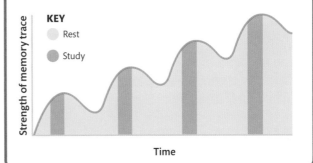

KEY
Rest
Study

Strength of memory trace

Time

Recalling a memory

Recalling a memory is not the passive process we once thought, like playing back a recording on your phone. Instead, our brain actively reconstructs our experience from the information it has stored. This introduces the opportunity for mistakes, meaning our memories can change over time.

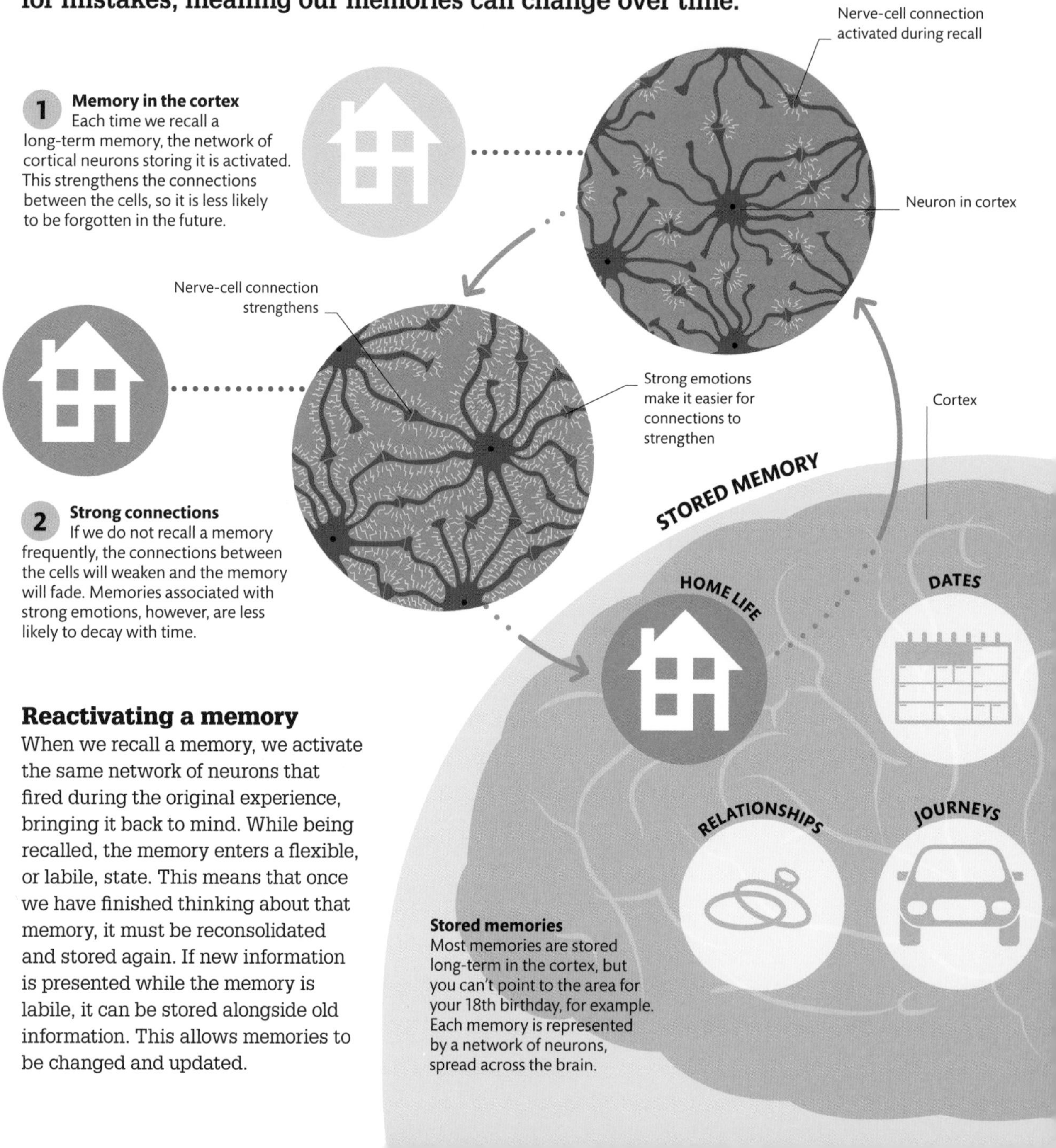

Nerve-cell connection activated during recall

1 Memory in the cortex
Each time we recall a long-term memory, the network of cortical neurons storing it is activated. This strengthens the connections between the cells, so it is less likely to be forgotten in the future.

Neuron in cortex

Nerve-cell connection strengthens

Strong emotions make it easier for connections to strengthen

Cortex

STORED MEMORY

2 Strong connections
If we do not recall a memory frequently, the connections between the cells will weaken and the memory will fade. Memories associated with strong emotions, however, are less likely to decay with time.

HOME LIFE

DATES

Reactivating a memory

When we recall a memory, we activate the same network of neurons that fired during the original experience, bringing it back to mind. While being recalled, the memory enters a flexible, or labile, state. This means that once we have finished thinking about that memory, it must be reconsolidated and stored again. If new information is presented while the memory is labile, it can be stored alongside old information. This allows memories to be changed and updated.

RELATIONSHIPS

JOURNEYS

Stored memories
Most memories are stored long-term in the cortex, but you can't point to the area for your 18th birthday, for example. Each memory is represented by a network of neurons, spread across the brain.

False memories

When a memory is reconsolidated, new information is stored with old. But when we next recall the memory, it is impossible to tell which is which. This means we can end up with false memories. Just talking about an event can change our memory of it, so in legal cases witnesses must be questioned carefully, to avoid contaminating their memories.

WHAT IS DÉJÀ VU?

The feeling of déjà vu might arise because we recognize something in an environment but cannot recall what. This gives a vague feeling of familiarity.

HOLIDAYS

BIRTHDAYS

1 True memory
Scientists asked participants to watch clips of car accidents. After each clip, they had to describe what happened and answer questions. This meant they were recalling and reactivating the memory.

2 New information
Some participants were asked about the cars' speed when they "contacted" each other, while others were asked about the speed when the cars "smashed". The first group rated the cars as slower than the second group.

TIME LATER

NEW INFORMATION STORED WITH OLD

3 False memory recalled
One week later, subjects recalled the video again, and were asked whether there was any broken glass (there was not). Significantly more people in the "smashed" group "remembered" broken glass. The words used had changed their memory of the event.

RECALL VERSUS RECOGNITION

It is much easier to recognize something as familiar when we are shown it than it is to recall the details without any input. For example, we all know what a pound coin looks like, but could you draw one from memory?

How to improve your memory

Once we understand learning and recall, research shows that we can find ways to boost these processes and improve our memories. Some of the best memory techniques, such as the memory palace, are actually some of the oldest.

Often, when we "forget" something, we haven't stored it properly in the first place. To avoid this, we must process information deeply – paying full attention to what we are learning, thinking about it, and seeing how it links to other things we already know.

Once stored, we need to make sure the information stays put, by practising or repeating whatever it is we are trying to learn. The more often we activate pairs of neurons together, the stronger that connection becomes and the more likely we are to remember it in the future. The spacing of repetitions is important too – it is better to revise for 10 minutes a day for six days than one hour on a single day.

The power of cues and rest

There are techniques we can use to help recall information, and many of them rely on cues. These triggers can be internal, such as mnemonics, which provide the first letters of a list of items, cueing recall of the items themselves.

Or they can be external – such as the scent of freesias taking you back to your wedding day. The memory palace technique uses associations and triggers to help recall long lists of information in order.

Probably the most important thing we can do for our memories is get enough sleep. If we are tired, our focus and attention suffer, and the brain just isn't in the right state to learn. Sleep is also vital after learning for memories to be consolidated, sorted, and stored.

Here is a quick recap of how to boost your memory:

- **Process the information deeply.**
- **Rehearse it regularly.**
- **Use cues and associations.**
- **Get plenty of sleep.**

Using a memory palace
Imagine you are walking through somewhere familiar, such as your house. At strategic points, visualize objects relating to the words you hope to remember, such as the items on a shopping list. To recall the list, simply "walk" the route again – the objects act as triggers.

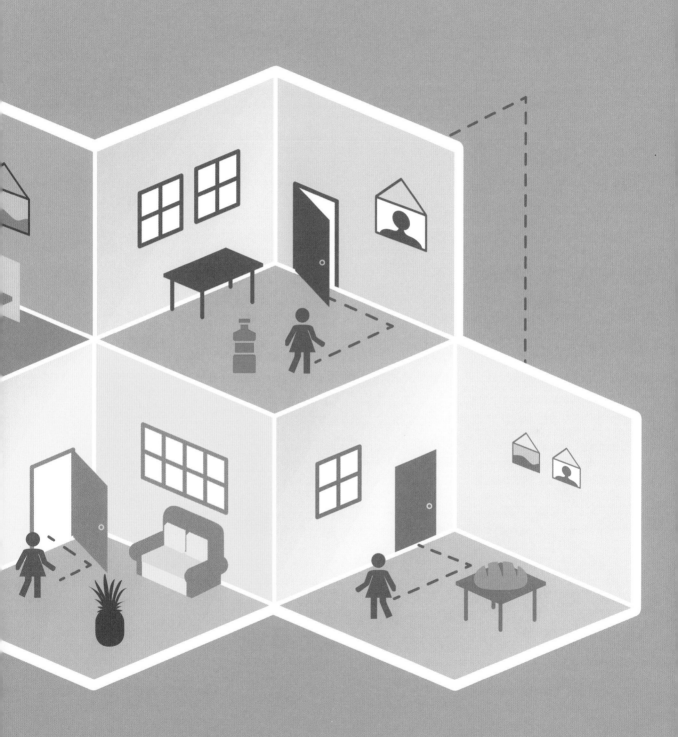

Why we forget

There are many theories to explain why we forget things. Some scientists think that all our memories remain in our brains but that we sometimes lose the ability to access them. Our memories may also interfere with one another.

Forgetting in the brain

There are many conditions that cause us to forget (see pp.146–47). Broadly, there are two possibilities for what happens in the brain when we do. The simplest idea is that over time memories fade away: information is lost as the trace that was formed is no longer there. But evidence for this is hard to come by, as other factors could be involved. Most of us have experienced the struggle to remember information that later pops into your head for no reason – this suggests memories can still exist but be inaccessible. This could be because other similar memories are interfering with them, or because there is no cue in our environment to prompt that recall. It is not known if the nerve-cell connections of a memory disappear, or whether they still exist but we are unable to access them.

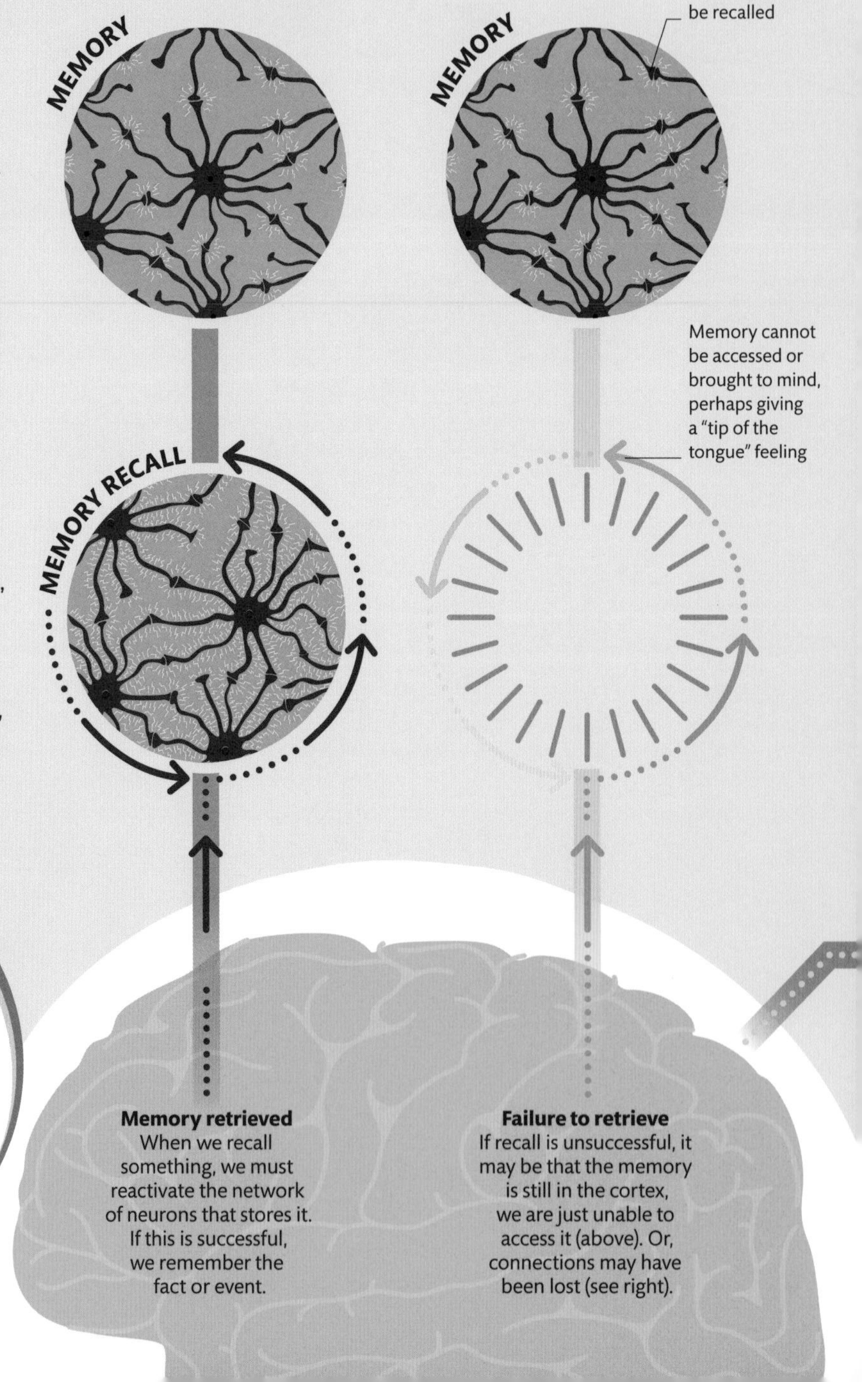

MEMORY

MEMORY

Memory trace exists in brain; often, blockage is later released and memory can be recalled

MEMORY RECALL

Memory cannot be accessed or brought to mind, perhaps giving a "tip of the tongue" feeling

WHY DO I FORGET WHAT I WENT UPSTAIRS FOR?

Leaving a room changes the environmental cues that help us remember. When you go back to where you were, the memory often reactivates.

Memory retrieved
When we recall something, we must reactivate the network of neurons that stores it. If this is successful, we remember the fact or event.

Failure to retrieve
If recall is unsuccessful, it may be that the memory is still in the cortex, we are just unable to access it (above). Or, connections may have been lost (see right).

Interfering memories

Our brains experience interference, particularly when information is similar. Learning new information can block recall for old, and old information can also affect new. These problems might arise because the wrong memory trace is activated when you go to recall the information, blocking access to the right one. Or, it may be that old information can disrupt consolidation of new, and if successful, the new memory may actually replace the old one.

ACTIVE FORGETTING

Forgetting seems passive, but you can choose to forget. In one study, subjects' prefrontal cortices – involved in suppression – were activated when they were told to forget a specific word.

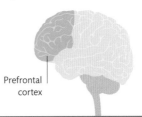

Prefrontal cortex

Proactive interference
Old memories may disrupt new ones. For example, when starting to learn Spanish, you may experience interference from French words learned as a child.

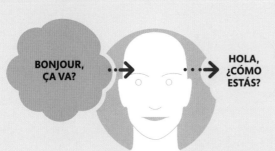

BONJOUR, ÇA VA?

HOLA, ¿CÓMO ESTÁS?

Retroactive interference
If you later went to speak French and instead spoke Spanish, that would be new memories disrupting the recall of old ones.

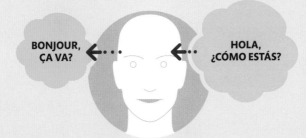

BONJOUR, ÇA VA?

HOLA, ¿CÓMO ESTÁS?

WE MAY BE **LESS LIKELY** TO **RECALL** INFORMATION WE CAN FIND EASILY **ONLINE**; THIS IS THE **GOOGLE EFFECT**

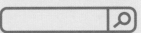

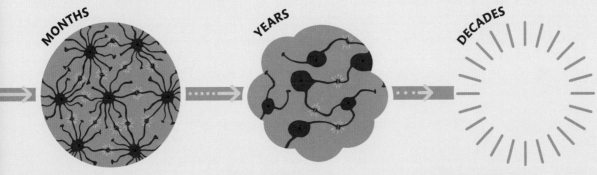

MONTHS

YEARS

DECADES

1 Storage
Long-term memories are stored in the cortex as networks of connections. These form and strengthen over weeks or months. Recalling a memory activates it, strengthening the synapses and making the memory easier to retrieve later.

2 Memory fades
If months or years pass before you recall a memory, it may begin to fade. Without reactivation, connections between nerve cells are not strengthened. Specific details about special events, such as the food you ate at your wedding, may be forgotten.

3 Losing a memory
One theory for forgetting is that synapses that are not in use become weaker, and are eventually pruned away, taking that memory with them. The longer a memory is inactive, the more likely it is to be lost through this process.

Memory problems

Memory problems increase with age, and dementia affects one in six people over 80. Sometimes, brain damage, stress, or other factors can cause us to experience an inability to remember (amnesia).

Amnesia

If someone suffers a brain injury that damages the hippocampus and surrounding areas, it can cause amnesia. There are two main types, depending on whether the patient forgets memories they had stored before the incident (retrograde amnesia) or is unable to form new memories (anterograde amnesia). There are also cases of amnesia without any obvious signs of damage, for example, after experiencing a psychological trauma. Drugs and alcohol can cause temporary amnesia, although this can become permanent if large amounts are used over a long period. It is also possible to suffer anterograde and retrograde amnesia at once, particularly if there is significant damage to the hippocampus. This condition is called global amnesia.

Retrograde amnesia
People often forget moments before an accident, but they can lose weeks, or even years. Some memories, especially older ones, return slowly.

Anterograde amnesia
People with anterograde amnesia are unable to form new memories. They remember who they are, and retain memories from before the damage.

Transient global amnesia
This is a sudden episode of memory loss, typically lasting a few hours. There are no other symptoms or obvious cause.

Infantile amnesia
Infantile amnesia refers to the fact that people usually cannot retrieve memories of situations or events before the age of two to four years.

Dissociative amnesia
This can be triggered by stress or psychological trauma. Patients forget days or weeks around the trauma or, in rare "fugue states", who they are.

Ageing and memory

As we age, it is normal to experience memory lapses and encounter more difficulty learning new things. Focusing attention and ignoring distractions becomes harder, and you may forget everyday things, such as why you went upstairs, more often. These experiences differ from the symptoms of dementia (see p.200), which can include getting lost in your own house or forgetting a partner's name.

BY THE TIME **PEOPLE REACH THEIR 80s** THEY MAY HAVE **LOST** AS MUCH AS **20 PER CENT** OF THE **NERVE CONNECTIONS** IN THEIR **HIPPOCAMPUS**

1 **Losing trust in memory**
Older adults often begin doubting their memories, seeing normal lapses as a sign of worsening abilities. This can lead them to rely on it less.

2 **Using memory less**
Brain abilities are like muscles, getting stronger with use. Writing things down or looking them up instead of exercising your memory could make it worse.

3 **Memory getting worse**
Not exercising your memory can cause a viscious cycle of cognitive decline. Encouraging older adults to use their memory, by providing feedback showing it still functions well, may help.

A curious case

Henry Molaison (1926–2008) was an American assembly line worker suffering from severe epileptic seizures. In 1953, he underwent surgery to remove sections of his medial temporal lobe, including both hippocampi, to treat severe epilepsy. This controlled his seizures, but he forgot several years before the surgery and developed anterograde amnesia. He could only retain new declarative memories (see p.135) for a few seconds, but could learn new skills.

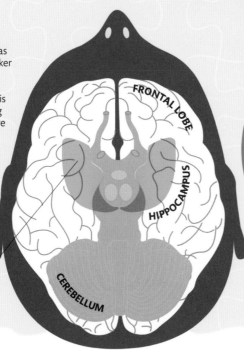

Large areas of medial temporal lobe removed from brain in each hemisphere

FRONTAL LOBE

HIPPOCAMPUS

CEREBELLUM

VIEW FROM BELOW

WHAT IS "SHELL SHOCK"?

The expression was coined during World War I to describe an effect thought to be caused by the sound of exploding shells. Soldiers were, in fact, suffering from PTSD, brought on by the trauma of war.

Other memory problems

Many things affect memory, from short-term stress to life events, such as having children. Memory changes can be linked to changes in our neurochemistry. For example, cortisol is released when we worry and hormones surge in a pregnant woman around the time of birth. Lifestyle changes such as sleep deprivation also play a role.

CAUSE	EXPLANATION
Stress	Moderate, short-term stress can make it easier to form memories, but it becomes harder to recall facts you have already learnt. This may explain why the feeling of "going blank" during an examination is so common.
Anxiety	Long-term, or chronic stress, such as is experienced by people with anxiety disorders, can damage the hippocampus and other memory structures of the brain, causing memory problems.
Depression	Depression can impact the short-term memory and cause people to have difficulty recalling details of events they have experienced. Healthy people tend to remember positives better than negatives. In depression, this is reversed.
"Baby brain"	Pregnant women may experience mild decline in a range of cognitive abilities, although these are likely to be noticeable only to the women themselves. After the baby is born, sleep deprivation can worsen memory problems.

POST TRAUMATIC STRESS DISORDER

Normally when we store memories, the emotion fades over time, so we recall past events without reliving them. In post traumatic stress disorder (PTSD), sufferers fail to dissociate memory from emotion, and intrusive memories bring the fear flooding back. These memories can be activated by sights or sounds, and often the patient is unaware of their triggers.

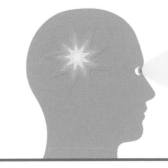

Special types of memory

Although a few children exhibit remarkable skills, most people with exceptional memory are not born that way. Instead, they use special techniques and lots of practice, sometimes leading to physical changes in their brains.

Training exceptional memories

Scientists studying trainee London taxi drivers as they learnt "the Knowledge" (a huge network of roads and landmarks) found that the volume of the subjects' posterior hippocampi increased as their ability to navigate improved. This could occur due to the birth of new neurons or the growth of existing dendrites (see p.20). However, the taxi drivers performed worse than control subjects in memory tests not involving London landmarks. This suggests memory is finite, and improving one area may come at the expense of others.

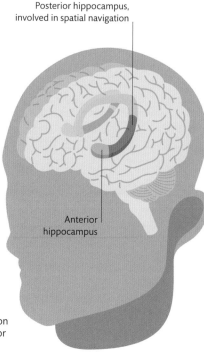

Posterior hippocampus, involved in spatial navigation

Anterior hippocampus

Hippocampal structures
Our two hippocampi – one on each side of the brain – are vital for learning and memory. They can be divided into posterior (back) and anterior (front), with the posterior portion particularly important for spatial navigation.

Savant syndrome

People with mental disabilities sometimes demonstrate incredible abilities in one specific area, often related to memory. This is called savant syndrome. Many savants are autistic, but the syndrome can also be triggered by severe head trauma. Some savants can calculate the day of the week for any given date. Others remember everything they read, or can paint detailed pictures of scenes they have only seen once. Scientists think these talents may develop because of savants' extreme focus and interest in one area. There is also evidence they see the world as building blocks, not whole pictures, by accessing perceptual information most of us are not consciously aware of.

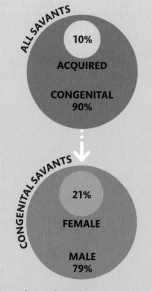

ALL SAVANTS

10%
ACQUIRED

CONGENITAL
90%

CONGENITAL SAVANTS

21%
FEMALE

MALE
79%

By genetics and gender
One database of savants, as reported by their parents or caregivers, found that the vast majority (90 per cent) are born with the condition, and of these, most were male.

FLASHBULB MEMORIES

People often remember where they were when receiving emotional news, and the memory seems extremely vivid and detailed. These are called flashbulb memories. However, studies have shown that we are as likely to be mistaken about these snapshots as we are about any other memories.

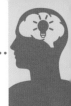

KEY
- Taxi driver's hippocampus
- Taxi driver's posterior hippocampus

Posterior hippocampus increases in volume

Posterior hippocampus returns to original size

Before training, taxi drivers have hippocampi with regions of normal size

1 Same size
At the start of the study, scientists scanned the brains of the participants to measure the size of their hippocampi. There were no differences between the trainee taxi drivers and the control group.

2 Changing anatomy
The trainee taxi drivers who passed the Knowledge had larger posterior hippocampi than the control group, or the trainees who failed. Some studies found that the front of their hippocampi was smaller.

3 Returning to normal
The brains of retired taxi drivers look much more like those of the control group. This suggests that the changes to the hippocampus revert once taxi drivers stop using the Knowledge on a daily basis.

"Photographic" memory

There is no such thing as photographic memory – no-one can literally recall pages of text or images as if they were really in front of them. The closest is eidetic memory, which occurs in 2–10 per cent of children. After looking at an image, "eidetikers" continue to "see" it in their visual field, until it gradually fades or disappears as they blink.

Picture imperfect
Studies have shown that eidetic images are not perfect. Children may not manage to remember all the letters in a word they were shown, or may invent details, for example "recalling" something in a picture that was not really there.

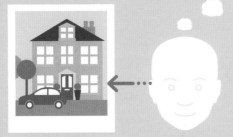

PHOTOGRAPH

CHILD

MEMORY
Sometimes, people with an eidetic memory vividly recall details that were not present in the original scene, such as the colour of this roof

CAN PEOPLE REMEMBER EVERYTHING?

A perfect memory does not exist, but a few people have superior autobiographical memory, giving them exceptional recall for events during their lives.

PEOPLE WITH INCREDIBLE RECALL FOR FACES ARE CALLED SUPER RECOGNIZERS

Intelligence

There are many theories about how intelligence evolved, what it actually constitutes, and which factors are key to high intelligence.

What is intelligence?

Intelligence is our ability to acquire information from our surroundings, incorporate that information into a knowledge-base, and then apply it to new situations and contexts. While there are many models for how human intelligence evolved, language and social living undoubtedly played a role as this enabled knowledge to be passed on from generation to generation. The evolution of human intelligence has led to our success as a species, enabling us to adapt to and inhabit almost all environments on Earth.

THERE ARE OVER **1,000 HUMAN GENES** THAT HAVE BEEN LINKED TO **INTELLIGENCE**

1 **Acquire** Information is gathered through various experiences, understood, and retained for processing.

2 **Process** New information is critically analysed, compared with existing knowledge and placed in context.

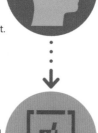

3 **Apply** Existing knowledge is applied to a new situation or problem, as opposed to being repeated from memory.

Network implicated in hypothesis testing – an integral component of intelligence

Frontal lobe houses large-scale networks associated with intelligence

Theories of intelligence

Some studies suggest that connectivity between the prefrontal and parietal cortices and small areas of neurons (networks) is the key to high intelligence (above). Other explanations (right) have also been put forward, suggesting that intelligence is related to connectivity across the brain as a whole.

Types of intelligence

Intelligence is often spoken of in a broad sense, but there is a theory that multiple intelligences exist. It recognizes that people may have the capacity to acquire and apply knowledge in specific areas. For example, someone may struggle with solving maths problems, but can reproduce a piece of music having only heard it once. Some argue this theory supports a more realistic definition of intelligence, while critics claim that these "intelligences" are merely aptitudes.

 Naturalist Recognizes features of plants and animals and infers insights based on what is known about the natural world.

 Musical Sensitive to rhythm, pitch, tone, melody, and timbre, and applies this to playing and composing music.

 Logical-mathematical Quick with numbers and easily quantifies things. Works out problems systematically and thinks critically about issues.

 Existential Uses observations, insight, and knowledge to explain the external world and the role of humans in it.

 Interpersonal Sensitive to people's moods, feelings, and motivations. Applies this to relationships and helping groups function.

Bodily-kinesthetic Uses heightened body awareness, coordination, and timing to master physical activities such as sport.

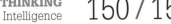

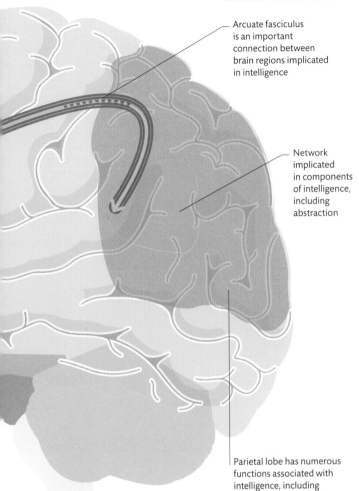

Arcuate fasciculus is an important connection between brain regions implicated in intelligence

Network implicated in components of intelligence, including abstraction

Parietal lobe has numerous functions associated with intelligence, including spatial awareness

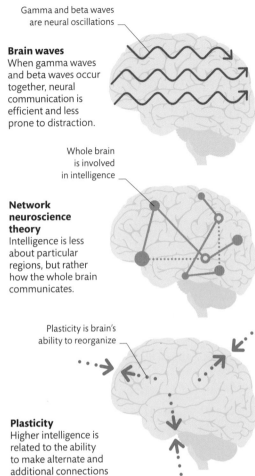

Gamma and beta waves are neural oscillations

Brain waves
When gamma waves and beta waves occur together, neural communication is efficient and less prone to distraction.

Whole brain is involved in intelligence

Network neuroscience theory
Intelligence is less about particular regions, but rather how the whole brain communicates.

Plasticity is brain's ability to reorganize

Plasticity
Higher intelligence is related to the ability to make alternate and additional connections within the brain.

Linguistic
Has a way with words, and uses this understanding to craft stories, convey complex concepts, and learn languages.

Intrapersonal
A deep understanding of self that can be used to predict one's own reactions and emotions to new situations.

Visual–spatial
Able to easily judge distance, recognise fine details, and solve spatial problems by visualizing the world in 3D.

INTELLIGENCE IS INHERITED

Physical features are not the only traits passed from one generation to the next. In fact, intelligence is thought to be one of the most heritable behavioural traits in humans. It is estimated that between 50 and 85 per cent of the differences in adult intelligence can be explained by genetics.

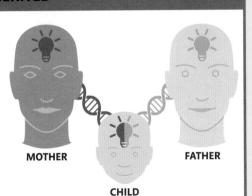

MOTHER

FATHER

CHILD

Measuring intelligence

Measures of intelligence have been used for well over a century, but the methods used and the way the results are put to use remain hotly debated, even today.

IQ scores are standardized so curve is always centred on a score of 100

Normal distribution
When scores from IQ tests are plotted on a frequency graph, the result is a bell curve, or normal distribution, in which most people's scores cluster symmetrically around the average. For every 100 people, 68 will have an IQ score between 85 and 115. At both the upper and lower ends of the scale, the frequency falls away rapidly.

AN INDIVIDUAL'S **IQ SCORE CAN VARY** BY **20 POINTS OR MORE** DEPENDING ON THE TEST USED

DOES A PERSON'S IQ STAY THE SAME?

A child's IQ score can be quite variable with potentially dramatic changes in score over relatively short periods of time. IQ scores tend to stabilize as adults.

IQ
Intelligence quotient (IQ) is a total score derived from a standardized test that measures aspects of intelligence, including analytical thinking and spatial recognition. There are over a dozen different tests that provide an IQ score, and they have been used to stream students and recruit to professions such as the military. Although IQ tests are statistically reliable, it has been argued that they are biased towards the cultures from which they originate.

Following a US court ruling in 2002, prisoners with an IQ lower than 70 cannot be considered for capital punishment

FREQUENCY

| 0.1% | 2.1% | 13.6% | 34.1% | 34.1% |

CATEGORY

55	70	85	100	115
LOWER EXTREME	WELL BELOW AVERAGE	LOW AVERAGE	AVERAGE	HIGH AVERAGE

IQ

Alternatives to IQ

IQ is not the only measure of intelligence. There are several alternatives, many of which are more visually based, with pictures, illusions, or pattern sequences at their core. Psychometric testing is an approach often used in job recruitment to assess a person's aptitudes – for example, to evaluate empathy when selecting a carer. People who score well on IQ tests are also likely to score well on other tests. This probably indicates a high level of overall cognitive ability, sometimes referred to as general intelligence factor (g).

General intelligence
The ability to do well across several specific areas of intelligence is indicated by the general intelligence factor.

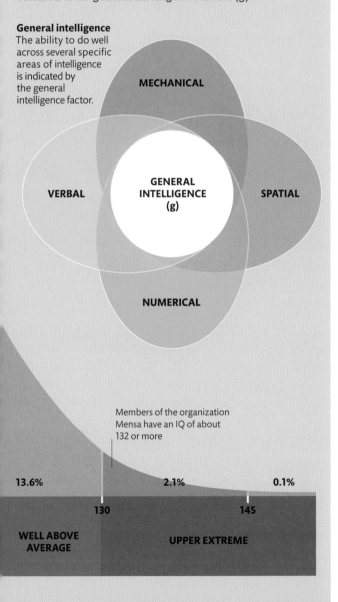

Members of the organization Mensa have an IQ of about 132 or more

13.6%   2.1%   0.1%

130   145

WELL ABOVE AVERAGE   UPPER EXTREME

RECORD IQS

Claims of exceptional IQs (including scores over 200) are often made but rarely verified. The American Marilyn vos Savant held the IQ record (228) in the Guinness World Records from 1986 to 1989, after which Guinness retired the category because it concluded the tests were not reliable enough. Attempts have also been made to measure the IQs of people who can no longer be tested. Albert Einstein, for example, is estimated to have had an IQ of over 160.

Is IQ on the rise?

There is evidence for a widespread increase in IQ. When IQ tests are revised every 10–20 years, the test-takers who are used to standardize the new test are asked to take the previous test as well, and they consistently score higher on the old test. In other words, if American adults today took an IQ test from the 1920s, the vast majority would score in the upper extreme, above 130. This is supported by evidence from around the world, although the rate of increase is most rapid in developing countries. Recent evidence suggests that this rise, known as the Flynn effect, has started to plateau.

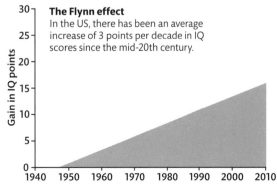

The Flynn effect
In the US, there has been an average increase of 3 points per decade in IQ scores since the mid-20th century.

Creativity

We all get a creative spark from time to time, but what makes some of us more creative than others is linked to our connections and coordination between three different brain networks.

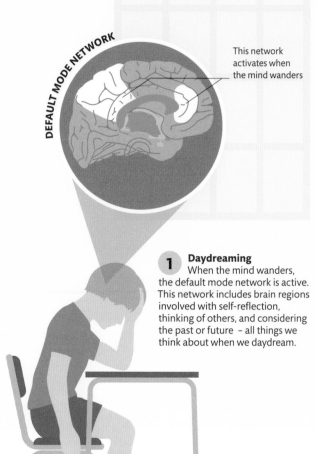

DEFAULT MODE NETWORK

This network activates when the mind wanders

The science of creativity

Creativity – our ability to come up with new and useful ideas – is linked to three distinct brain networks: the default mode network, the salience network, and the central executive network. While these networks are linked, they are not typically active at the same time. However, fMRI studies of people asked to perform specific tasks show that people who can switch quickly between these networks at suitable moments have more creative responses to the task. The correlation is so strong, in fact, that a person's creativity can be predicted based on the strength of the connection between these networks.

1 Daydreaming
When the mind wanders, the default mode network is active. This network includes brain regions involved with self-reflection, thinking of others, and considering the past or future – all things we think about when we daydream.

JAPANESE INVENTOR **SHUNPEI YAMAZAKI** HAS A REPORTED **5,255 PATENTS** TO HIS NAME

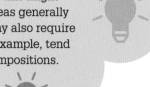

The creative brain

While genetics plays a role in creativity, other factors are also significant. Low levels of noradrenaline may support creativity as this neurotransmitter diverts inward-focused attention to external stimuli. While this might help our fight-or-flight response, creative ideas generally emerge from internal sources. Creativity may also require a strong knowledge base – composers, for example, tend to write their best work after decades of compositions.

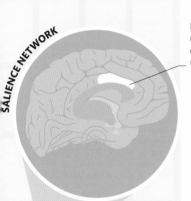

Recruits other networks based on information received

SALIENCE NETWORK

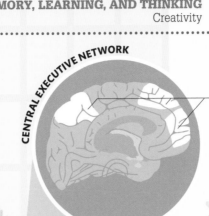

CENTRAL EXECUTIVE NETWORK

Regions activated to maintain attention on particular task

2 Switching
The salience network detects sensory information to determine if the central executive network should engage. For example, when hearing your name while daydreaming, the salience network will trigger a switch.

3 Focusing
The central executive network engages the conscious brain to think and maintain focus on a task. Studies have shown that the default mode network is re-engaged within a fraction of a second of the task being completed.

THE BRAIN ON JAZZ

In one study, jazz musicians were asked to play the piano while in an fMRI machine. Their brain activity was recorded as they switched from playing memorized music to improvised jazz. The results showed that brain areas responsible for the evaluation of our own actions and inhibition were less active during improvisation.

Activity in the lateral prefrontal cortex

Deactivation in lateral prefrontal cortex

MEMORIZED MUSIC

IMPROVISED MUSIC

WHY DO IDEAS OFTEN FLOW WHEN WE ARE NOT FOCUSED ON A TASK?

The brain is particularly good at reconfiguring and connecting information when it is not in a task-orientated mode.

How to boost your creativity

Just as exercise builds muscles and improves cardiovascular fitness, there are activities that can improve your creative conditioning, by getting areas of the brain to work together in new ways.

To boost creativity, you must first remove barriers to it. Stress, time constraints, and lack of sleep or exercise are known creativity-killers. People tend to be creative when they are rested, happy, and can let their thoughts wander freely. Many people claim to have their best ideas during their morning shower or walk to work. It seems that ideas flow most freely around our brains when they are not in a task-orientated state but instead in a condition called the resting state.

Cultivate new connections

Routines help regulate our daily lives, but they also reinforce existing neural pathways.

Creativity-friendly activities create new neural connections. Learning to play a musical instrument, for example, opens and strengthens links between different brain areas.

Simply varying your routine can also foster creativity, so pick a more interesting route to work, a colour you don't usually wear, or a new recipe to cook. Surround yourself with like-minded, creative people as much as possible. Whether it is in a gallery or a garden shed, new input stimulates new ideas.

Unsolvable challenges encourage novel ways of thought. How many things can you think of to do with a paperclip, for instance? If you are stuck on a problem, get some

mental distance from it. Imagine how someone from another country, time period, or age group would deal with the issue.

Allow yourself to disconnect. If you are stuck in a queue, don't default to your phone to check emails or social media; instead, zone out and let the ideas flow.

The next time you are stuck for ideas, try one of the following:
- **Get enough rest, de-stress, and exercise.**
- **Learn a new skill. Spend time with other creative people.**
- **Think outside the box. Think of new ways to solve old problems.**
- **Switch off from digital devices to give your brain some down-time.**

Belief

Our brains can distil complex information, taking unexplainable observations and evaluating and categorizing them. From this we form propositions – true or not – that guide us in life.

How do our beliefs form?

Our beliefs develop out of what we hear, see, and experience, from our interactions with others and with our environment. They are entwined with our emotions, which is why an emotional response is often evoked when those beliefs are challenged. Beliefs are accepted as truth, whether there is proof or not. Our beliefs then become a filter, where information that does not support those beliefs is rejected, potentially limiting our perceptions of the world. Beliefs are not static though – each of us has the power to choose and change our beliefs.

Knowledge
What you know impacts on beliefs and challenges those held.

Future vision
How you imagine life to be is intricately linked to your beliefs.

Facets of belief
We process information from many aspects of life in order to form our beliefs. Equally, our beliefs also shape how we process this information.

Events
Positive and negative events both shape how you view the world.

Environment
Where, how, and who raised you underpins many beliefs.

Past results
Successes and failures shape your beliefs about what is possible.

Ventromedial prefrontal cortex activated in belief

Insula registers disbelief

1 Bad behaviour
The human brain is exceptional at spotting patterns in even random phenomena. Before humans understood what lightning was, for example, they looked for patterns, and many cultures around the world believed it coincided with bad behaviour.

2 Brain areas
Regions of the brain involved in emotions are important in establishing beliefs. The biochemical basis of beliefs is an active area of research as evidence, including the placebo effect, suggests that beliefs trigger biochemical responses in the body.

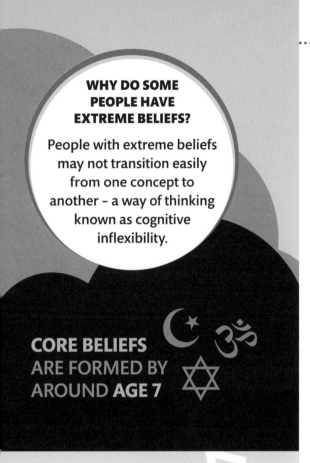

WHY DO SOME PEOPLE HAVE EXTREME BELIEFS?

People with extreme beliefs may not transition easily from one concept to another – a way of thinking known as cognitive inflexibility.

CORE BELIEFS ARE FORMED BY AROUND AGE 7

The layers of belief

The deepest layer of beliefs, core beliefs, are the principles that guide our actions (processes). It is our actions that then determine what our outcomes are. When we are looking to make changes in our life, we often focus on outcomes as these are the easiest to change in the short-term. However, to foster long-lasting change, we need to change our habits, and to do this we may need to examine our core beliefs.

Core beliefs
Your core beliefs are intertwined with how you view yourself and the world around you and are therefore the most tightly held and inflexible.

OUTCOMES

PROCESS

CORE BELIEF

REASONING BELIEFS

There are three types of beliefs: factual, preference, and ideology. If two people are debating factual beliefs, only one of them can be right, whereas both people can be right in the case of preference. Ideological beliefs draw elements from both fact and preference. Pre-school children can differentiate between these types of beliefs and recognize that in some cases two people can be right.

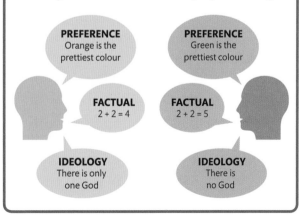

PREFERENCE
Orange is the prettiest colour

PREFERENCE
Green is the prettiest colour

FACTUAL
2 + 2 = 4

FACTUAL
2 + 2 = 5

IDEOLOGY
There is only one God

IDEOLOGY
There is no God

3 Supernatural explanation
As well as spotting patterns, the human brain favours intention over randomness. The idea that lightning was intentionally wielded by gods to punish bad behaviour was therefore more satisfying than it being a random natural event.

CONSCIOUSNESS

AND THE SELF

What is consciousness?

Consciousness is our awareness of external stimuli (such as our surroundings) and internal events (such as our thoughts and feelings). We can identify the brain activity that generates conscious awareness, but how this phenomenon arises from a physical organ remains a mystery.

Locating consciousness

Our thoughts, feelings, and ideas are all activities of the brain – products with a neurological basis. However, it is unclear whether it is the neurological activity itself that forms consciousness (or the "mind") or whether it is merely linked to consciousness. This is the fundamental difference between two theories of consciousness. The first, monism, equates the mind with the brain, while the second, dualism, sees the mind as separate from the brain and body.

LIGHT

MONISM

DUALISM

Monism
According to monism, every thought, feeling, and idea is a product of the brain activity that occurs as the result of a stimulus. This brain activity is itself the conscious perception of the object. In other words, the brain is the mind and vice versa.

Where is the mind?
When we see an object, it is the result of our brain perceiving a light stimulus. However, whether this activity in our brain directly leads to consciousness, or whether the activity links to an external mind, is debated.

VIRTUAL REALITY

Virtual and augmented realities are no longer restricted to the plots of science fiction. Computers are now used to simulate external stimuli – such as sights or sounds – that provide the brain with an alternative reality.

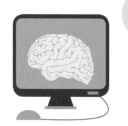

BRAINSTEM DEATH

In parts of the world (such as the UK), the legal definition of death is brainstem death. Irreversible damage to the brainstem (see p.36) prevents it from regulating the automatic functions essential to life. These may be continued with the help of medical equipment, but the person will never regain consciousness.

Dualism
The dualist theory argues that the mind (which is non-physical) exists outside of the brain (which is physical), but that the two interact. The brain activity that happens as a result of the stimulus is associated with conscious perception, but the mind itself is separate.

COULD ARTIFICIAL INTELLIGENCE BECOME CONSCIOUS?

Some scientists do believe that artificial intelligence could be programmed to be conscious; others believe consciousness is not something that machines could ever learn.

The requirements of consciousness

The neural basis of consciousness remains an area of research, which is directed at identifying the structures and processes in the brain that are necessary to produce a conscious experience. The process of consciousness is believed to be at the level of individual neurons, rather than at the level of individual molecules or atoms. It is likely that for consciousness to arise, the four factors below must be present.

HIGH FIRING RATES

BETA BRAINWAVES

A normal state of consciousness occurs when neurons fire at fairly high rates. Beta waves (see p.42) occur when neurons fire at a high rate, and indicate alertness and logical, analytical thinking.

SYNCHRONOUS FIRING

Consciousness may depend on the synchronicity of neurons. Clusters of neurons firing in unison "bind" individual perceptions – such as sight, sound, and smell – to create one perception.

IN **1 OR 2** OF EVERY **1,000 MEDICAL PROCEDURES** INVOLVING **GENERAL ANAESTHESIA** A PATIENT MAY BECOME **CONSCIOUS**

TIMING

It can take half a second for the unconscious brain to process stimuli into conscious perceptions – but our brain is capable of making us think that we experience things immediately.

FRONTAL ACTIVITY

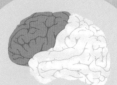

The frontal lobes may play an important role in aspects of consciousness, including feelings of reflection, as well as coordinating levels of consciousness.

Paying attention

Attention directs our consciousness (see pp.162–63) to focus more intently on a particular sensory input, such as a sight or sound, and to tune out competing information. The process of paying attention begins with the sensory organs, which activate various areas of the brain, including the frontal and parietal lobes. The parietal lobe processes spatial information, directing attention to an area of space, while the frontal lobe directs the eyes to focus on specific objects.

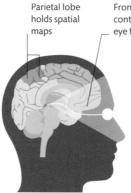

Parietal lobe holds spatial maps

Frontal lobe contains frontal eye field

OPTIC NERVE

Superior colliculus acts as a tracking system, directing head and eyes to follow an object

Attention areas
Key to paying attention to visual stimuli is the frontal eye field, located in the frontal lobe, and the superior colliculus. Together, they instruct our eyes to focus on an object.

Attention

Attention is the process of concentrating or focusing on specific information. The brain is the main organ that processes both behavioural and cognitive information, although other parts of the body, such as the eyes and ears, are also required.

RESEARCH SUGGESTS THAT THE **AVERAGE** HUMAN **ATTENTION SPAN** IS JUST **8 SECONDS**

ATTENTION DEFICIT HYPERACTIVITY DISORDER

Attention deficit hyperactivity disorder (ADHD) is a behavioural disorder (see p.216) that includes symptoms such as inattentiveness and hyperactivity. The exact cause of ADHD is not yet fully understood. Research suggests that there may be an imbalance of neurotransmitters or a genetic cause. Any potential genetic cause of ADHD, however, is thought to be complex and is unlikely to be caused by a single gene.

ARE OUR ATTENTION SPANS SHRINKING?

There is no evidence that our individual attention spans are shrinking, but a recent study suggests that our collective attention span – how long as a society we focus on a news story or trending topic, for example – is decreasing.

SUSTAINED ATTENTION

Sustained attention is the ability to concentrate on a specific task, such as reading a book, for a long period of time. Brain imaging studies have shown that the frontal and parietal cortical areas, particularly in the right hemisphere of the brain, are associated with sustained attention.

SELECTIVE ATTENTION

Selective attention is the process of focusing intently on something specific, such as an object or sound, while tuning out our environment. Ignoring the sound of a car while paying attention to a phone is an example of selective attention.

Types of attention

There are various types of attention, and the sort of attention that is required depends upon the circumstances that we are in. Both sustained and selective attention are used when we need to focus fully on one stimulus. Alternating and divided attention are used when there are multiple inputs that we need to focus on at the same time. Attention is not an unlimited resource and the process of focusing our attention on something can be tiring, as it needs a significant amount of energy.

ALTERNATING ATTENTION

Alternating attention is the ability to switch attention quickly between tasks that require a very different cognitive response. Cooking dinner while periodically checking a recipe in a book is an example of alternating attention between different tasks.

DIVIDED ATTENTION

Divided attention is used so that we are able to perform two or more activities at the same time; for example, riding a bicycle while listening to music. This type of attention is sometimes called multitasking.

Distractions

The brain is not able to focus our attention constantly. Instead, it cycles rapidly between two different states: attention and distraction. During periods of distraction, the brain scans the environment to check that there is nothing more important to which it should be paying attention. This cycle is thought to give an evolutionary advantage to humans, allowing us to respond quickly to either new opportunities or threats.

During periods of distraction, brain scans environment

Looking for trouble
Even when we think we are focused on a task, our brain is checking the environment so that attention can be diverted if necessary.

How to focus your attention

Focusing your attention requires your brain to process specific information. Learning how to accomplish this in a world where there are plenty of distractions is crucial to allow you to learn, understand, and function properly.

Attention is a limited resource, and focusing it has to be carefully managed if you are to be able to limit distractions and concentrate on specific tasks. The ability to focus attention varies between people. It is influenced by both your interest in the task at hand and the number of distractions that you encounter. If you are really interested in something, you may not even notice other distractions that occur in the environment around you. This is simply because it is easier to focus your attention on something if you are engaged with it. So, how can you increase your ability to focus your attention?

Distractions, distractions, distractions

Focusing your attention involves concentrating on something specific, while tuning out both external and internal distractions. While you are reading this book, you will hopefully be focusing your attention on the words written in the text. However, your brain will be bombarded with a range of distractions. These can emanate from a variety of external sources. For example, the television may be on in the background or there may be people having a conversation around you.

You might also be faced with internal distractions. Hunger may motivate you to start thinking about what you are having for dinner. You might suddenly remember an important task that had slipped your mind. These types of internal thoughts are driven by an area of the brain called the medial prefrontal cortex (see pp.30–31), which is associated with decision-making, emotional

responses, and the retrieval of long-term memories.

Research suggests that once your attention is distracted from completing a task, it can take an average of 25 minutes to return to the original exercise. So, the next time you are being distracted, try one of the following to focus your attention:

▪ **Keep potential distractions away. Turn off any electronic devices and move to a quiet place.**
▪ **If the task at hand is unavoidably monotonous, it can help to remind yourself why you are doing it.**
▪ **Imagine the sense of accomplishment you will feel upon completing the task. This can provide additional motivation.**
▪ **Gradually and slowly increase the time that you try to focus your attention. This can improve your attentional focus.**

Free will and the unconscious

Many activities in everyday life – from our movements to our emotions – are not controlled consciously. Instead, unconscious activity in the brain is behind a lot of our actions, thoughts, and behaviours.

Free will

The ability to choose a course of action without restriction is called free will, and it may seem that we use our conscious mind to make these decisions. However, research suggests that we may have less conscious control over our actions than we think. Experiments have shown that our brain begins to plan a movement one-fifth of a second before we consciously decide to make a move.

CAN YOUR UNCONSCIOUS HELP YOU SOLVE A PROBLEM?

If you are stuck on a problem, letting your mind wander can allow the brain to collect information from your unconscious and potentially provide a solution.

Benjamin Libet's experiment
Scientist Benjamin Libet instructed his subjects to note down when they became conscious of their decision to raise a finger. At the same time, their brainwaves and muscle movements were recorded.

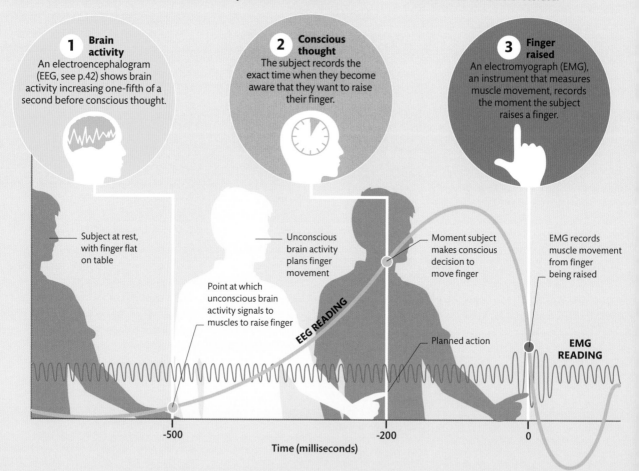

1 Brain activity
An electroencephalogram (EEG, see p.42) shows brain activity increasing one-fifth of a second before conscious thought.

2 Conscious thought
The subject records the exact time when they become aware that they want to raise their finger.

3 Finger raised
An electromyograph (EMG), an instrument that measures muscle movement, records the moment the subject raises a finger.

Subject at rest, with finger flat on table

Unconscious brain activity plans finger movement

Point at which unconscious brain activity signals to muscles to raise finger

EEG READING

Moment subject makes conscious decision to move finger

EMG records muscle movement from finger being raised

Planned action

EMG READING

-500   -200   0

Time (milliseconds)

Levels of consciousness

In the early 20th century, neurologist Sigmund Freud popularized the idea that the mind is divided into three levels of consciousness: the conscious mind (mental processes we are aware of), the preconscious (processes we are not aware of but can be brought into the conscious), and the unconscious (inaccessible mental processes that influence our behaviour). More modern thinking suggests that there are several levels of consciousness, ranging from intense self-reflection to the deepest sleep.

Introspection
We examine our thoughts, actions, and emotions – for example, we may brood over an action we have taken.

Normal consciousness
We have a sense of agency – we believe that we control our thoughts and that they influence what we do.

Unconscious knowledge
We can perform complex tasks though may not have a memory of doing so – for example, not remembering the drive home.

Lack of consciousness
Asleep, we neither perceive the world around us nor have the sense of self to experience things such as time passing.

95 PER CENT
OF OUR **DECISIONS** ARE **MADE** BY OUR **UNCONSCIOUS MIND**

Ironic process theory

If we are asked not to think of a white bear, we will probably think of a white bear. This is because a deliberate attempt to suppress a thought makes it more likely to occur. This phenomenon is explained by an idea known as ironic process theory. The idea is that the brain unconsciously monitors itself for occurrences of the unwanted thought – which, ironically, then makes us aware of the thought. This is partly why quitting smoking is difficult or why trying to forget a bad memory rarely works – the unconscious reminds us of the things we are trying to forget.

In 2006, two Dutch researchers asked subjects to make a complex decision under one of three conditions: with little time for consideration; with ample time; or with ample time but distractions that prevented conscious thought about the decision. In all cases, the distracted subjects performed best. The findings suggest that people can make better decisions unconsciously than consciously – although the experiment suggested this is true only when we are making complicated decisions.

Altered states

An altered state of consciousness is any condition that differs significantly from our normal state of consciousness (see pp.162–63). It is almost always temporary and always reversible.

Physical and physiological
Extreme environmental conditions, such as high altitudes or weaker gravity in space, can induce altered states, as can extended fasting and breath manipulation.

Types of altered state
Altered states can be grouped into categories based on how they are induced. However, all altered states disrupt brain function in some way.

Psychological
An altered state can be induced through certain cultural or religious practices, such as meditation or trances brought on through dancing or drumming. Other examples are sensory deprivation and hypnosis.

Spontaneous
Spontaneously induced altered states include drowsiness, daydreaming, near-death experiences, and the state of consciousness that happens just before you fall asleep (known as a hypnagogic state).

Disease-induced
Disease and illness can alter the conscious experience to different degrees. Examples include psychotic disorders such as schizophrenia (see p.211), as well as epileptic seizures and coma.

Pharmacological
Psychoactive (mind-altering) drugs, such as alcohol, cannabis, or opioids, disrupt how the brain's neurotransmitters function, altering the user's awareness and consciousness levels.

IS A NEAR-DEATH EXPERIENCE AN ALTERED STATE?

This is highly debated, but those who have had such experiences describe elements, such as a sense of timelessness, common to other altered states.

What is an altered state?
When we are in a normal state of consciousness, we are aware of external stimuli (such as our surroundings) and internal events (such as our thoughts). However, the brain can produce a much wider range of conscious experiences, including altered states. Whenever we enter an altered state, our brain patterns change. This disruption in brain function can be caused in different ways, including changes in blood flow and oxygen to the brain or interference with neurotransmitter function.

Controlled and automatic processes

The way we are able to perform controlled processes (tasks that require our full awareness, such as solving a puzzle) and automatic processes (tasks that require relatively little attention, such as reading a book) is compromised.

Self-control

We may have difficulty controlling our actions and movements, for example walking a straight line while intoxicated. It may also be difficult to restrain emotions, often resulting in outbursts of crying or aggression.

Level of awareness

In an altered state, our level of awareness of events going on around us – as well as internally – may be increased or decreased compared with normal waking consciousness. More often, our level of awareness is lowered in an altered state.

Identifying an altered state

Consciousness is a spectrum from highly alert to total lack of awareness, with a "normal" state somewhere in the middle. Altered states, meanwhile, can be on either side of the scale, with greater or lesser awareness than normal. An altered state can be identified using different criteria.

Emotional awareness

Often in an altered state we will have less emotional awareness (the experience of emotions), as well as finding it difficult to control those emotions. This can make us more or less affectionate, aggressive, or anxious.

Perceptual and cognitive distortions

Perception may be altered. Normal processes for storing and retrieving memories may be more fragmented or less accurate. Thought processes may be disorganized and less logical.

Time orientation

In an altered state, our sense of time (see pp.174–75) can become distorted; time may appear to slow down or speed up. This is because there is less awareness of time passing, just as we are unaware of time while we sleep.

382 DAYS
THE LONGEST RECORDED FAST FROM SOLID FOOD

Altered states in the brain

Altered states can lead to a range of experiences, from feelings of bliss to a sense of terror. These experiences are generated by a similarly diverse range of neural activity in various parts of the brain. Alterations to normal brain function can result in our brain distorting incoming information, leading to auditory or visual hallucinations, memory distortion, or delusions.

Decrease in activity in frontal lobe reduces ability to reason and make decisions

Thalamus – which acts as gateway between limbic system and frontal cortex – can be inhibited

Altered activity in parietal lobe distorts spatial judgements and time perception

Changes in temporal lobe function lead to unexplainable experiences such as hallucinations

Signals from reticular formation, which plays important role in consciousness, can be reduced

Locating altered states

In an altered state, activity in different areas of the brain may increase or decrease, distorting how we perceive the world.

Sleep and dreams

When we are asleep it may seem like our brains are quietly resting, but they are actually busy processing and storing information that we have learned throughout the day.

The stages of sleep

During the night, we cycle through different sleep stages, moving from light to deep sleep, then to rapid-eye-movement (REM) sleep. Our brain waves, produced by the electrical activity of neurons in the cortex (see p.42), change in each stage. As sleep becomes deeper, the waves become slower (with lower frequency) and more organized. We repeat this sleep cycle every few hours, but the proportions shift; we have more slow-wave sleep at the start of the night and more REM sleep in the early morning.

(see p.42)

A not-so-silent night

There are four distinct stages of sleep, and we pass through each stage several times a night. During light sleep, we are easily woken. It is much harder to wake from deep sleep.

HOW MANY HOURS OF SLEEP DO WE NEED A NIGHT?

Most adults need 7–9 hours of sleep a night, but teenagers and children (especially babies) need more.

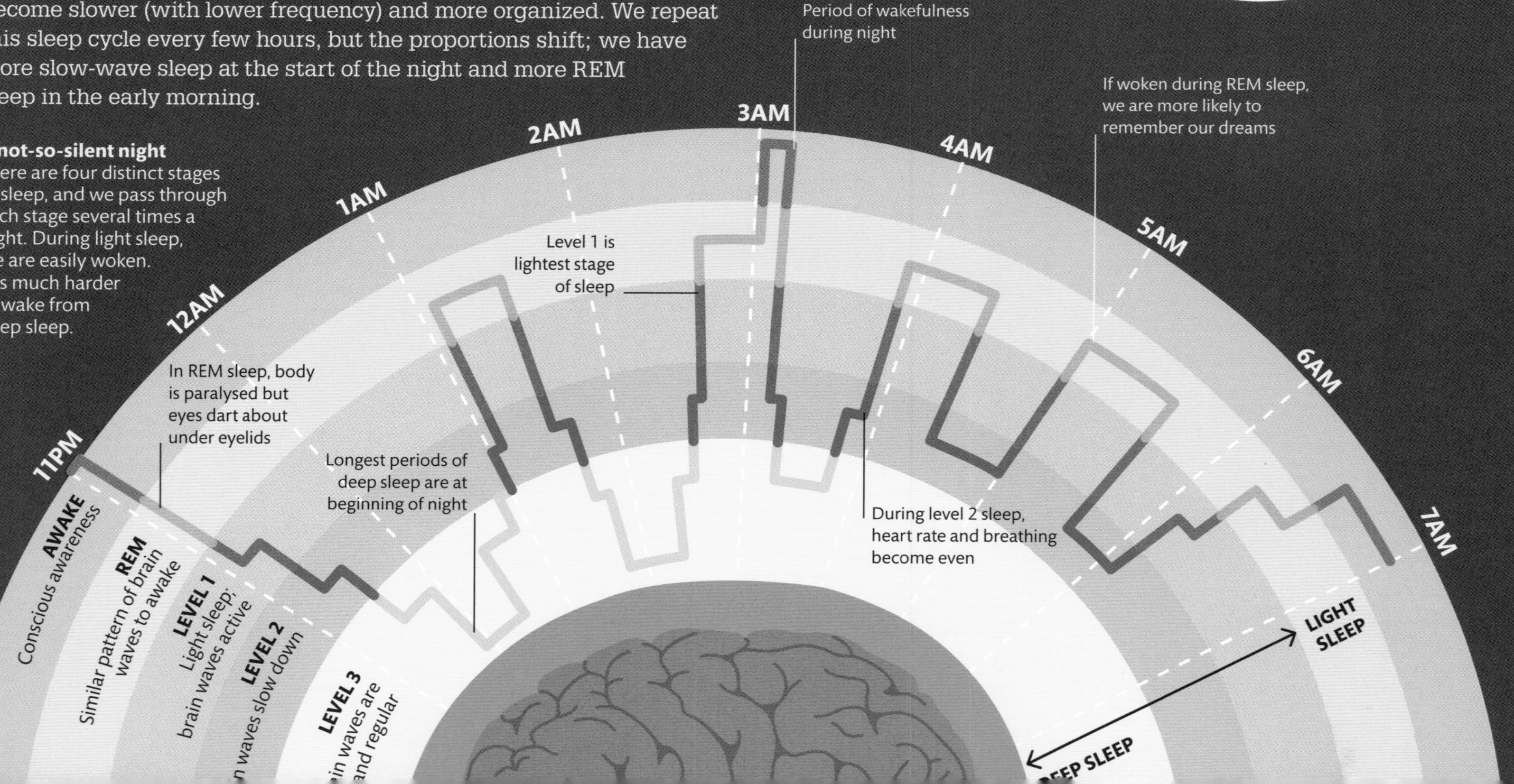

Period of wakefulness during night

If woken during REM sleep, we are more likely to remember our dreams

During level 2 sleep, heart rate and breathing become even

Level 1 is lightest stage of sleep

Longest periods of deep sleep are at beginning of night

In REM sleep, body is paralysed but eyes dart about under eyelids

LIGHT SLEEP

EEP SLEEP

7AM
6AM
5AM
4AM
3AM
2AM
1AM
12AM
11PM

AWAKE
Conscious awareness

REM
Similar pattern of brain waves to awake

LEVEL 1
Light sleep: brain waves active

LEVEL 2
n waves slow down

LEVEL 3
in waves are and regular

THE GLYMPHATIC SYSTEM

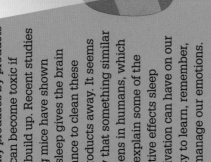

LYMPHATIC DUCT

Debris swept away by cerebrospinal fluid

Astrocytes shrink, allowing fluid through

Neurons produce debris

BLOOD VESSEL

Flow of cerebrospinal fluid

There is evidence to suggest that while we sleep, some of our brain cells shrink, allowing cerebrospinal fluid to flow more easily between them. The fluid carries away any waste that has accumulated to the lymphatic ducts, where it is removed from the body.

Cleaning the brain

During the day, our brain activity produces by-products that can become toxic if they build up. Recent studies using mice have shown that sleep gives the brain a chance to clean these by-products away. It seems likely that something similar happens in humans, which may explain some of the negative effects sleep deprivation can have on our ability to learn, remember, and manage our emotions.

THE **LONGEST** RECORDED ATTEMPT TO **STAY AWAKE** IS **264 HOURS**

SLEEP DISORDERS

Problems like sleepwalking, sleep talking, and paralysis occur when the brain fails to make a clean shift between sleep states. This leaves part of our brain awake while other parts are sound asleep. When a person sleepwalks, the motor areas of the brain are awake and active but the conscious awareness and memory areas are asleep. People can even perform complex tasks such as driving while fast asleep.

Thalamus delivers signals to cortex

Parietal cortex, which controls awareness of oneself, is inactive

Hippocampus sends new memories to cortex

Visual cortex generates imagery

Reticular formation switches between sleep and wakefulness

Areas of prefrontal cortex are inactive, so reason not applied to dreams

Amygdala generates emotions

KEY

Active

Inactive

Activity during REM sleep
Emotional brain regions are very active during REM sleep, as is much of the cortex. The frontal lobes, involved in rational thinking, are much less active.

The dreaming brain

Scientists do not know why we dream, but they have theories. Dreams might help us to process information and emotions encountered during the day and store them in our long-term memory (see pp.138–39). A dream might also be like a rehearsal – our brain is trying out responses to extreme events in safety so we would be prepared if the event happened in real life. This might explain why dreams are often stressful or negative. Another idea is that dreams are merely "screensavers" for the mind, with no real purpose at all.

Time

We can measure time objectively – by hours, minutes, and seconds – with clocks, but our brain also helps us to keep track of time passing. Our internal clocks are all set at different speeds and even change within our lifetime.

The timekeeper brain

Our concept of time is linked to a neural network involved in memory and attention. Neurons in the network fire, or "oscillate", and the brain uses this to keep time. The more oscillations in a measured second, the more we think that time is lasting longer. Events (such as near-death experiences), state of mind (such as depression), stimulants (such as caffeine), and disease (such as Parkinson's disease) can all affect the rate at which the neurons fire, skewing our perception of time.

Direction of dopamine flow

Anterior part of prefrontal cortex

Basal ganglia

Substantia nigra

The dopamine clock

Another one of the brain's clocks is formed of the oscillation, or cycle, of dopamine flowing between the substantia nigra, basal ganglia, and prefrontal cortex.

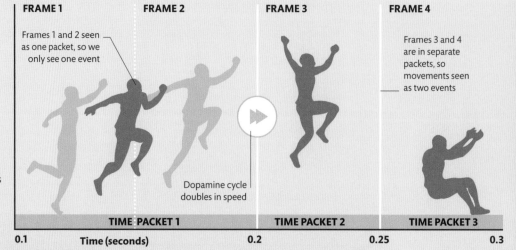

FRAME 1 **FRAME 2** **FRAME 3** **FRAME 4**

Frames 1 and 2 seen as one packet, so we only see one event

Frames 3 and 4 are in separate packets, so movements seen as two events

Dopamine cycle doubles in speed

TIME PACKET 1 TIME PACKET 2 TIME PACKET 3

0.1 Time (seconds) 0.2 0.25 0.3

Packets of time

One cycle of a brain clock equals one "packet" of time, which we register as a single event. Just as a camera with a higher frame rate will capture more details in a sequence of events, faster rates of neuronal firing will create more time packets, registering more events.

TIME ILLUSIONS

Distance can skew our perception of time. If three lights flash one after another at equal time intervals (of 10 seconds, for example), but the distance between light "B" and "C" is greater than the distance between "A" and "B", it will create the illusion that the time between "B" and "C" flashing was longer than 10 seconds.

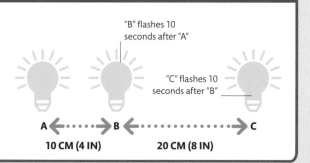

"B" flashes 10 seconds after "A"

"C" flashes 10 seconds after "B"

A B C

10 CM (4 IN) 20 CM (8 IN)

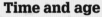

Time and age

It can feel like time speeds up as we get older – a journey that felt like an eternity as a child passes quickly as an adult. Part of the reason for this is that our perception of time develops as we age. As infants, we live in the moment – we cry if we are not fed on time, but we are not aware of the passage of time. As toddlers, we are taught to become aware of time, and we learn how long it takes to perform everyday tasks, such as brushing our teeth. By the time we are six years old, we can estimate time, applying our knowledge of how long something takes to new situations.

Factors affecting time perception

As adults, we are more conscious of time, as we have responsibilities and schedules. These routines of moving from one event to the next can speed up our perception of time. However, there are also biological, proportional, and perceptual theories as to why time seems to speed up with age.

HOW DO DRUGS AFFECT TIME PERCEPTION?

Dopamine is the main neurotransmitter involved in time processing. Some drugs, such as methamphetamines, activate dopamine receptors, speeding up the perception of time.

Metabolism

In a 24-hour period, a four-year-old's heart will have done 125 per cent of the beats of an adult heart. Other biological markers, such as breathing, are also faster. This means children take in more information, so time appears to move slowly.

PERCEPTION OF TIME IS SUSPENDED WHEN WE ARE ASLEEP ZZ^Z

Proportional theory

As we age, time intervals constitute smaller fractions of our lives as a whole. For example, one year is 10 per cent of a 10-year-old's life, but only 2 per cent of a 50-year-old's life.

Perceptual theory

The more information we absorb and process, the slower we perceive time to be. Children, who are experiencing many things for the first time, pay more attention to details that adults dismiss, which may stretch out time.

Pathways in the brain

As we age, the pathways in our brain grow more complex, so signals take longer to travel along them. This means older people view fewer images in the same amount of objective time, so time seems to pass more quickly.

What is personality?

Our personality makes us who we are. It is a set of behavioural characteristics that shape the choices we make in life and how we react to the world. Various systems have been invented to assess and classify personality.

Changeable personality

From the moment we are conceived, DNA begins to shape our personality, leading us to produce more of a certain neurotransmitter than another, for example, or making us less sensitive to a hormone compared to other people. This affects our underlying temperament, and even our final personality to some extent. However, as well as our genetics, who we are is also shaped by our experiences and environment.

Becoming you

As we grow, our brains mature along set patterns and change through experience. Regularly used neural pathways become stronger, and we may become more or less reactive to neurotransmitters and hormones. This changes our personality.

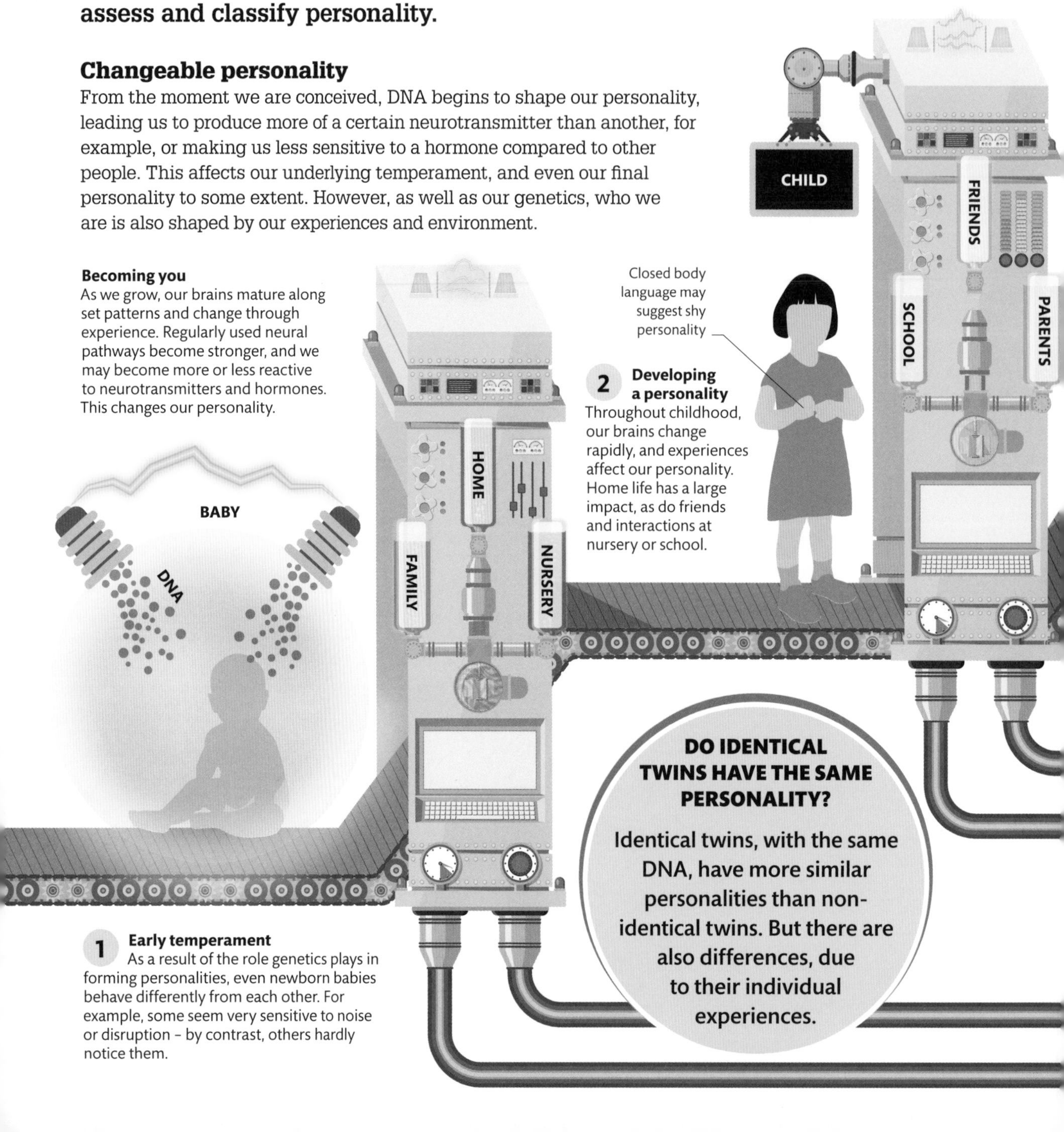

BABY

DNA

CHILD

FRIENDS

SCHOOL

PARENTS

Closed body language may suggest shy personality

2 **Developing a personality**
Throughout childhood, our brains change rapidly, and experiences affect our personality. Home life has a large impact, as do friends and interactions at nursery or school.

HOME

FAMILY

NURSERY

1 **Early temperament**
As a result of the role genetics plays in forming personalities, even newborn babies behave differently from each other. For example, some seem very sensitive to noise or disruption – by contrast, others hardly notice them.

DO IDENTICAL TWINS HAVE THE SAME PERSONALITY?

Identical twins, with the same DNA, have more similar personalities than non-identical twins. But there are also differences, due to their individual experiences.

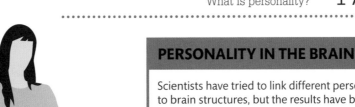

PERSONALITY IN THE BRAIN

Scientists have tried to link different personality types to brain structures, but the results have been mixed. We do know that brain damage, particularly to frontal areas, can have an impact on someone's personality, and studies have linked certain traits to differences in brain structure or activity. So far, however, the complexities of both the human brain and our behaviour have made the links hard to unravel.

Crossed arms may indicate defensiveness or insecurity

Fashion choices used to reflect personality

ADULT

3 **Adult personality**
As well as environmental factors such as school or friends, our personality alters due to the fact that our brains do not finish maturing until our early twenties. Our personality goes on to change subtly throughout adulthood.

Assessing personality
The most common personality assessment, the Big Five test, identifies how a person scores in terms of five traits: openness, conscientiousness, extroversion, agreeableness, and neuroticism. A person is placed along scales for each trait, with one end being the least likely to exhibit this trait, and the other the most.

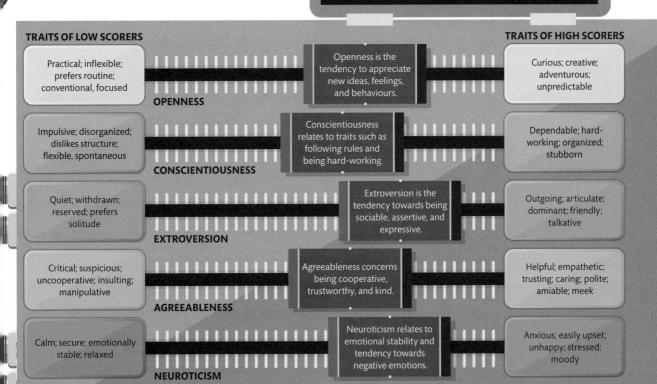

TRAITS OF LOW SCORERS

TRAITS OF HIGH SCORERS

Practical; inflexible; prefers routine; conventional, focused	Openness is the tendency to appreciate new ideas, feelings, and behaviours.	Curious; creative; adventurous; unpredictable
OPENNESS		
Impulsive; disorganized; dislikes structure; flexible, spontaneous	Conscientiousness relates to traits such as following rules and being hard-working.	Dependable; hard-working; organized; stubborn
CONSCIENTIOUSNESS		
Quiet; withdrawn; reserved; prefers solitude	Extroversion is the tendency towards being sociable, assertive, and expressive.	Outgoing; articulate; dominant; friendly; talkative
EXTROVERSION		
Critical; suspicious; uncooperative; insulting; manipulative	Agreeableness concerns being cooperative, trustworthy, and kind.	Helpful; empathetic; trusting; caring; polite; amiable; meek
AGREEABLENESS		
Calm; secure; emotionally stable; relaxed	Neuroticism relates to emotional stability and tendency towards negative emotions.	Anxious; easily upset; unhappy; stressed; moody
NEUROTICISM		

The self

The self is an accumulation of concepts of who we are, who we were, and who we want to be. We derive our sense of self in different ways, through awareness of ourselves as physical beings, as agents of our actions, and as a part of society.

What is the self?

The self is our internal sense of who we are, which develops through our evaluation of our experiences of the world. It is formed of two aspects: the physical self (who we are as tangible beings) and a mental self (which can be seen as our autobiographical memory). There are several linked areas of the brain that contribute to our sense of self. Our physical sense of self is created by areas that tell us how our body occupies space, while areas that allow us to reflect on our mental state and retrieve memories contribute to our mental self.

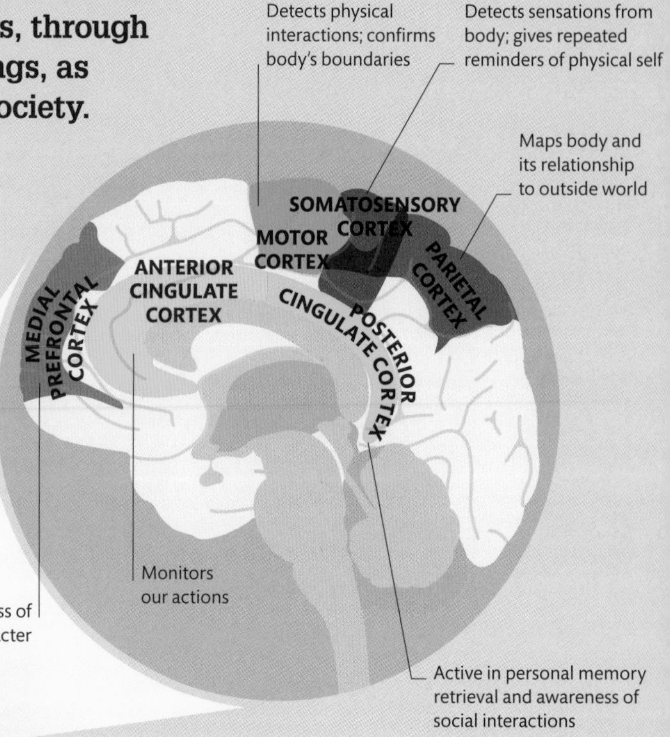

Detects physical interactions; confirms body's boundaries

Detects sensations from body; gives repeated reminders of physical self

Maps body and its relationship to outside world

SOMATOSENSORY CORTEX

MOTOR CORTEX

PARIETAL CORTEX

ANTERIOR CINGULATE CORTEX

MEDIAL PREFRONTAL CORTEX

POSTERIOR CINGULATE CORTEX

Monitors our actions

Enables consciousness of mental state and character

Active in personal memory retrieval and awareness of social interactions

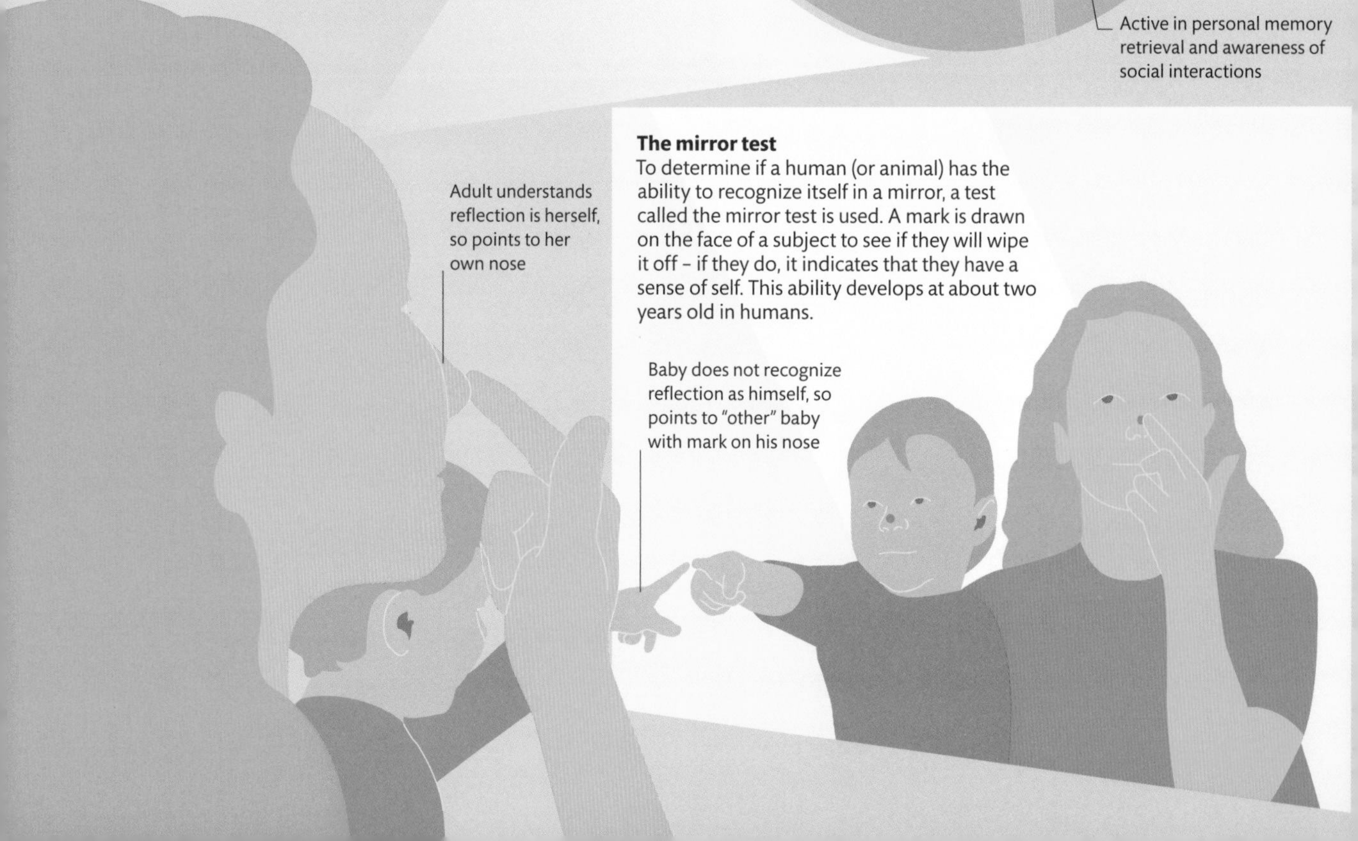

Adult understands reflection is herself, so points to her own nose

The mirror test
To determine if a human (or animal) has the ability to recognize itself in a mirror, a test called the mirror test is used. A mark is drawn on the face of a subject to see if they will wipe it off – if they do, it indicates that they have a sense of self. This ability develops at about two years old in humans.

Baby does not recognize reflection as himself, so points to "other" baby with mark on his nose

The actual and ideal self

There can sometimes be a difference between who we believe we are (our actual self) and who we aspire to be (our ideal self). How we perceive our actual self shifts in response to feedback and challenges from the social environment. Some psychologists believe that when our actual self is close to our ideal self we are more able to live a balanced, happy life.

Congruence
When the difference between our actual self and ideal self is small, we are said to be "congruent".

Small overlap indicates our actual self does not reflect who we aspire to be

Large overlap suggests our actual self is similar to who we aspire to be

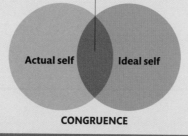

Actual self   Ideal self

Actual self   Ideal self

INCONGRUENCE   **CONGRUENCE**

SELF AND IDENTITY

The self is a first-person account of how we perceive and evaluate ourself. Identity involves the specific beliefs and characteristics that can be used to define a person and distinguish them from others.

The development of self

The concept of self begins as soon as we are able to recognize that we are an individual being that is distinct from other objects and people. This basic sense of self happens shortly after birth, but it is not until our second year of life that we begin to develop a more complicated view of who we are.

Am I liked?

I am good.

I am 3.

DO DOGS RECOGNIZE THEMSELVES IN A MIRROR?

Dogs fail the mirror test, but some scientists have argued that the test might not work for animals that do not rely on sight as their primary sense.

2 YEARS OLD   **3–4 YEARS OLD**   **6 YEARS OLD**

60 PER CENT OF SOCIAL MEDIA USERS SAY IT NEGATIVELY IMPACTS HOW THEY FEEL ABOUT THEMSELVES

Self-description
By two years old, toddlers begin to refer to themselves as "me". They often describe themselves as they may be perceived by other people.

Categorical sense of self
Young children define themselves in terms of properties and categories – these are usually concrete, such as age or hair colour.

Defining self against peers
By school-age, children start comparing themselves to their peers. Many beliefs about their self stem from how others react to them.

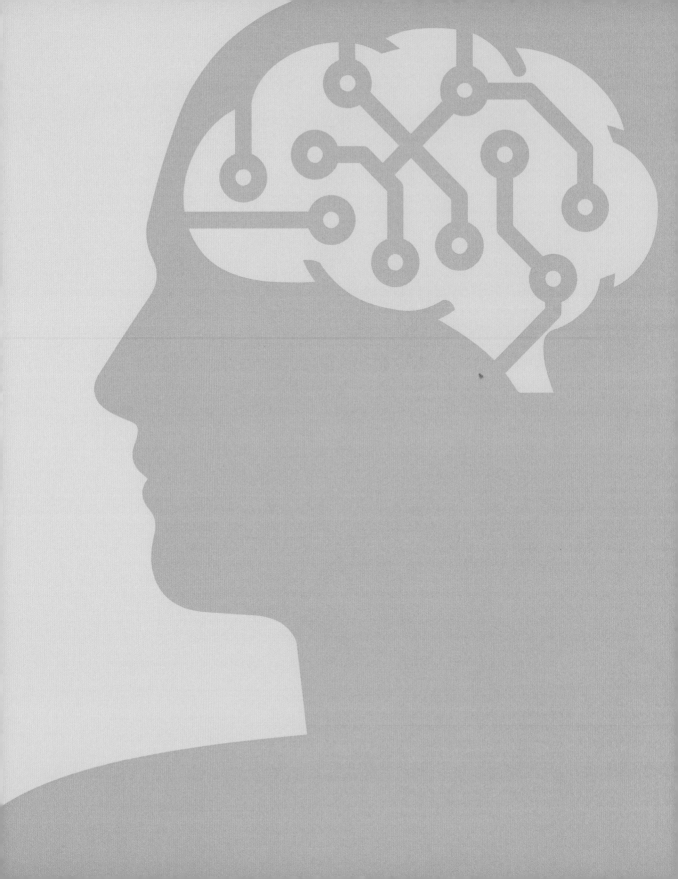

THE BRAIN OF
THE FUTURE

Superhuman senses

The latest electronic devices almost rival our eyes and other sense organs. Future versions may not only restore lost sensory function but even expand our range of sensations.

Transmitting sight and sound

Cochlear implants were introduced in the 1970s and retinal implants first appeared in 2011 to help people with severe hearing and sight problems respectively. Video cameras and microphones "catch" light and sound and convert them into signals that travel to a processing unit. This creates a digital "map", which is relayed via wireless signals to an implant. The implant sends the data via nerve impulses to our relevant sensory region of the brain.

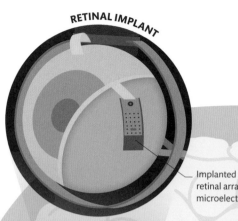

RETINAL IMPLANT

Implanted retinal array of microelectrodes

3 Data transmitted to implant
The relay sends wireless signals to the antenna of a prosthesis on the side of the eyeball. The antenna passes the signals along wires to a retinal array implanted inside the eye.

SOMATOSENSORY CORTEX

Electrodes stimulate olfactory bulb

VIDEO CAMERA

AUDITORY CORTEX

Camera captures images

Wire travels to electrodes implanted in nostril

ANTENNA

RETINAL IMPLANT

WIRE CONNECTS TO ELECTRODE

Electrosniffers
Some "electronic noses" feature copied human proteins that work as receptors, creating electric pulses that travel along a wire when contacted by a certain substance.

Optic nerve carries impulses from deeper retinal cells to visual cortex

Relay transmitter sends signals wirelessly to antenna on eyeball

1 Video camera
One or two small video cameras worn on spectacles form images from incoming light rays. The images are converted to electrical signals and sent along wires to a portable video processing unit (VPU).

2 Video data
The smartphone-sized VPU, worn on the body but potentially implantable, converts the camera's video signals into a digital "map" of spots or pixels. It sends this along wires to a receiver-transmitter relay mounted on the spectacles.

Airborne odour and flavour molecules enter nasal cavity

ESP

Some people report that they receive information or awareness that could not have originated from known sensory inputs. Such occurrences can be labelled extra-sensory perception (ESP) but can usually be explained by sudden recall of forgotten experience or coincidence. Future research may also reveal natural human abilities to detect magnetic fields and other phenomena.

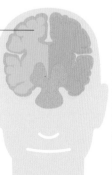

Scans show more right hemisphere activity in reported ESP

4 **Implant sends data to brain**
The retinal array is an electronic grid that sends signals to the deeper layers of cells in the retina, bypassing its faulty light-detecting cells. These deeper cells create nerve impulses that travel to the visual cortex.

ARTIFICIAL SKIN

Evolving forms of artificial skin contain graphene sheets with domed electronic sensors. Physical changes such as temperature and pressure stretch or squash these sensors to generate electrical signals that are then transmitted to the somatosensory cortex in the brain.

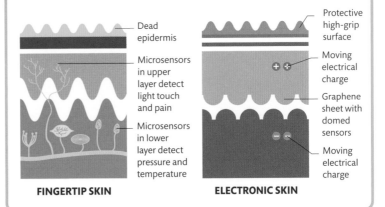

Dead epidermis

Microsensors in upper layer detect light touch and pain

Microsensors in lower layer detect pressure and temperature

FINGERTIP SKIN

Protective high-grip surface

Moving electrical charge

Graphene sheet with domed sensors

Moving electrical charge

ELECTRONIC SKIN

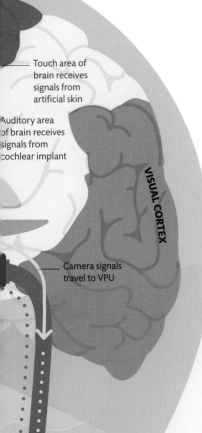

Touch area of brain receives signals from artificial skin

Auditory area of brain receives signals from cochlear implant

VISUAL CORTEX

Camera signals travel to VPU

Signals travel along wires from body-worn VPU

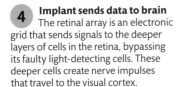

ELECTROSNIFFERS
DETECT SCENTS WITH AROUND **97 PER CENT** ACCURACY

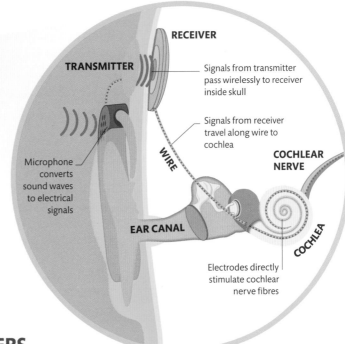

RECEIVER

TRANSMITTER

Signals from transmitter pass wirelessly to receiver inside skull

Signals from receiver travel along wire to cochlea

COCHLEAR NERVE

Microphone converts sound waves to electrical signals

WIRE

EAR CANAL

COCHLEA

Electrodes directly stimulate cochlear nerve fibres

Cochlear implant
Many designs of cochlear implant bypass damaged parts of the outer and middle ear and the sensory cells of the inner ear's cochlea. They work by supplying tiny electrical signals directly to cochlear nerve fibres.

Wiring the brain

Until recently, only the brain controlled the body's muscles and glands. But next-generation electrical, mechanical, and robotic devices – often developed after limb loss – are extending its abilities.

1 Motor cortex
The brain's movement centre formulates patterns of motor nerve impulses that naturally coordinate dozens of muscles to move the arm and hand.

Somatosensory cortex — Motor cortex

Spinal cord links to arm nerves

Bionic limbs
Motorized bionic limbs now exist that react to activity in the brain's motor cortex, responding to instructions sent as tiny electrical impulses along motor nerves. These increasingly powerful prostheses can also provide sensory feedback so that the brain's control systems can provide delicate ongoing control, more closely mimicking the natural limb or other body part.

Wires carry digital signals to servos in hand

2 Sending impulses
Motor nerve impulses travel from the brain via the spinal cord along peripheral nerves to the arm and hand.

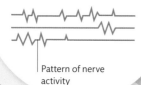

Pattern of nerve activity

3 Microprocessor
Microchips change the nerve impulses into digital signals understood by the circuits and motors of the bionic part.

Impulses converted to digital signals

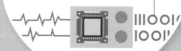

4 Bionic hand
Up to 10 servos (small, lightweight motors) drive movements of the hand and fingers, pivoting at self-sensing joints.

Hand receives processed signals and converts them to movement

Two-way communication
The motor cortex masterminds movements of the bionic part. As with a natural limb, these are continually modified by interchange with the somatosensory cortex.

Median, radial, and ulnar nerves

Motor impulses to bionic hand

6 Mindful awareness
Further processing converts the sensory signals to more natural forms that can be interpreted by the brain's touch centre, the somatosensory cortex.

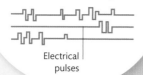

Electrical pulses

5 Sensory data
Receptors in the hand's motors, joints, and artificial skin generate responses.

IOIIIOOIOIOOIIO
OIIOOIIIOOIOIOI
OIIOOIOIOOIIIOI

Feedback signals produced by robotic hand in digital form

Sensory signals from bionic hand

Deep brain stimulation (DBS)

In DBS, electrode wires are implanted in various parts of the brain (see below) to treat a range of disorders. These send pulses of electricity from a generator and battery in the chest, connected to the electrodes. A remote controller adjusts the pulses. In adaptive DBS, the electrodes have sensors and the generator automatically responds to the brain's electrical activity.

THE **BATTERIES** USED IN PULSE GENERATORS FOR **DEEP BRAIN STIMULATION** LAST UP TO ABOUT **NINE YEARS**

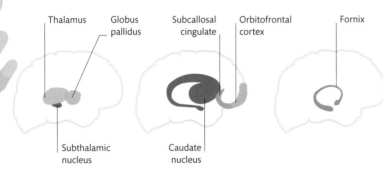

Thalamus Globus pallidus Subcallosal cingulate Orbitofrontal cortex Fornix

Subthalamic nucleus Caudate nucleus

Movement disorders
DBS is well-established to treat movement problems, such as the tremors and "freezing" of Parkinson's disease and the spasms and contractions of dystonia.

Psychiatric disorders
DBS may be used in severe anxiety, depression, and obsessive-compulsive disorder, where other treatments such as drug medication have not proved effective.

Cognitive disorders
Research explores DBS for problems such as Alzheimer's disease, targeting specific structures involved in memory and cognitive neural networks.

WHEN WAS THE FIRST BIONIC LIMB CREATED?

In 1993, a team of bio-engineers at the Margaret Rose Hospital in Edinburgh created the first bionic arm for amputee Robert Campbell Aird.

Vagus nerve stimulation

The vagus nerve, one of the cranial nerves (see p.12) connects the brain with organs in the chest and abdomen. In vagus nerve stimulation (VNS), a small signal generator in the chest, similar to a heart pacemaker, is connected by wires to electrodes around the left vagus nerve in the neck. The nerve's sensory fibres are stimulated to send impulses into the brain, where they are distributed along various neural pathways. VNS is mainly used to treat forms of epilepsy and depression.

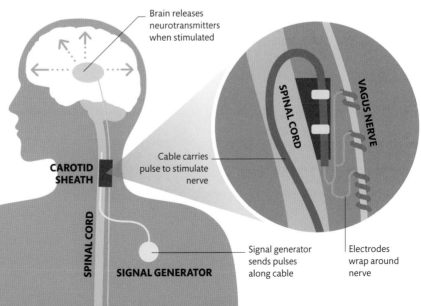

Brain releases neurotransmitters when stimulated

CAROTID SHEATH

SPINAL CORD

SIGNAL GENERATOR

SPINAL CORD

VAGUS NERVE

Cable carries pulse to stimulate nerve

Signal generator sends pulses along cable

Electrodes wrap around nerve

The unexplored brain

New research is revealing that some well-known parts of the brain have unexpected functions. This is especially true of the "lower brain" areas, such as the brainstem and thalamus – areas once thought to be largely passive and to perform only automated roles.

Discovering potential

Cutting-edge scanning methods can probe areas of the brain beneath the cortex to understand their contributions to conscious thoughts and behaviours. These techniques include magnetoencephalography (MEG), which detects magnetic fields generated by neurons (see p.43), and fMRI and near-infrared spectroscopy (NIRS), which monitor brain activity by detecting changes in local blood flow and oxygenation.

The brainstem and emotion

Far from being a routine life-support region, the brainstem (see pp.36–37) is active in our behaviour, especially emotions. Moods and feelings are even being localized to specific nuclei (clusters of nerve cells). These areas may be manipulated by electrodes or chemicals to treat problems such as depression, anxiety, and panic attacks.

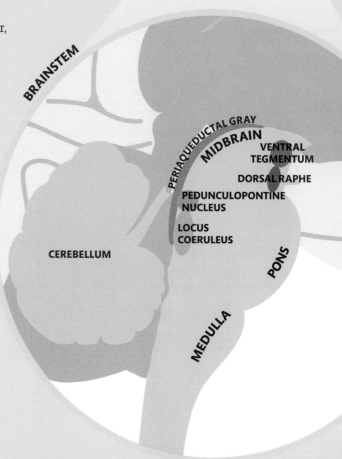

Dorsal raphe
This nucleus is a major source of serotonin. Problems here can lead to worry, anxiety, and low mood.

Locus coeruleus
Malfunction of this major producer of noradrenaline may cause intense emotions, stress, and poor memory.

Pedunculopontine nucleus
This has roles in focused attention and concentration, as well as in physical tasks, such as moving limbs.

Periaqueductal gray
Wrapped around the cerebral aqueduct channel, this nucleus is a major part of the pain-coping system.

Ventral tegmentum
This nucleus has a central function in motivation, learning, and reward, and is implicated in conditions such as ADHD.

THALAMUS

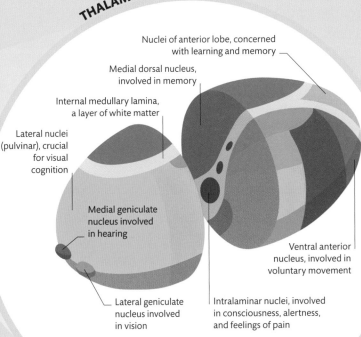

Nuclei of anterior lobe, concerned with learning and memory

Medial dorsal nucleus, involved in memory

Internal medullary lamina, a layer of white matter

Lateral nuclei (pulvinar), crucial for visual cognition

Medial geniculate nucleus involved in hearing

Lateral geniculate nucleus involved in vision

Ventral anterior nucleus, involved in voluntary movement

Intralaminar nuclei, involved in consciousness, alertness, and feelings of pain

Thalamic nuclei
Investigations into lesser-known nuclei are revealing lots of surprises. For example, the pulvinar nucleus helps the vision centres map out and measure a scene and how we reach out to objects there.

The brain's relay station

It is already well known that the thalamus acts as a relay station for all incoming sensory information (except smell), but more is now being discovered about how it pre-processes this information in a complex and selective manner before it travels to sensory zones in the cortex. The thalamus is also central for the regulation of arousal and, with its links to the hippocampus, it plays an important role in memory. Deep brain stimulation (see p.185) of the thalamus has been used to treat conditions including tremors.

DESPITE ITS **BODYWIDE EFFECTS**, THE **SCN** CONTAINS ONLY **20,000 NEURONS** AND IS SMALLER THAN THIS LETTER **O**

HAVE ALL THE BRAIN'S PARTS BEEN DISCOVERED?

Not yet. In 2018, improved microscopes uncovered a small region at the brain–spinal cord junction, which was named the endorestiform nucleus.

THE SCN

Located in the hypothalamus, the tiny suprachiasmatic nucleus (SCN) sets the body's circadian rhythm – our 24-hour sleep-wake cycle. This biological clock drives vital homeostatic functions including body temperature, feeding, and hormone levels. The SCN also coordinates the activities of many organs. Microscopic electrodes or lasers could one day adjust these cycles and patterns.

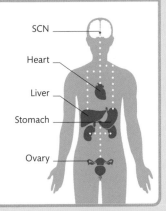

SCN

Heart

Liver

Stomach

Ovary

Artificial intelligence

As computers become more sophisticated, the ultimate goal is to develop a machine that passes the Turing Test, in which a person in conversation with the machine cannot tell that they are not talking to another person.

Delivering dropout
Many electronic neural networks analyse and process in stages. In dropout, the probability is assessed that a particular item of information will or will not be useful. If it is not, it is removed.

Mimicking the brain

Computer programs called neural networks attempt to copy the way the brain works by using artificial neurons arranged in layers. Inspired by the way people learn, neural networks can adapt and change their responses over time (see right), a feature known as machine learning. To more closely replicate the human brain's highly adaptive, generalized intelligence, a more advanced approach involves querying, modifying, and deleting data, a technique called adaptive forgetting. For example, data that is little used further along a network, as shown by feedback through the system, can be trimmed or deleted. This is called dropout. Reducing this redundant data produces a system that is faster and more compact and responsive.

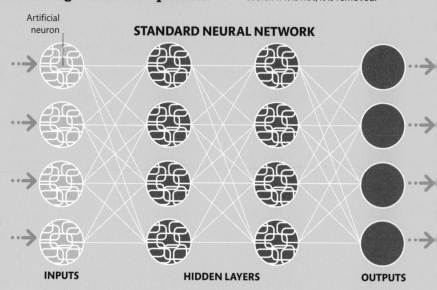

Artificial neuron

STANDARD NEURAL NETWORK

INPUTS   HIDDEN LAYERS   OUTPUTS

1 Input layer
The network receives inputs in the form of numbers, or values. For example, in an image-recognition system, an input might be the brightness of an individual pixel in a digital image.

2 Hidden layers
The hidden layers process the data they receive from the input layer. Over time, the network "learns," modifying its results by applying different weights to the values.

3 Output layer
Once it has been processed, data passes to the output layer. In the image-recognition system, the output would be the application's "guess" for what the image shows.

WILL ROBOTS TAKE OVER THE WORLD?

An "AI takeover" sounds like science-fiction but it is hypothetically possible. A lot depends on friendly computers preventing self-evolving ones from advancing beyond humans.

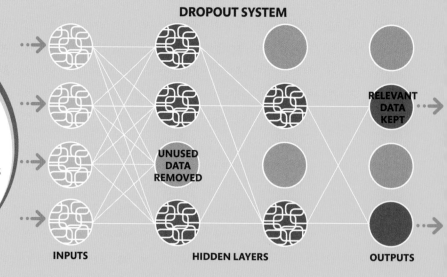

DROPOUT SYSTEM

INPUTS   HIDDEN LAYERS   OUTPUTS

RELEVANT DATA KEPT

UNUSED DATA REMOVED

Forming memory circuits

Modelling digital electronic circuits on the brain means storing and recalling information. In the brain, remembering involves the repeated use of particular pathways between neurons that strengthen their junctions (synapses) to form a "memory circuit". In electronics, a device in development known as the memory resistor or memristor offers a similar function.

KEY

Large resistance

Small resistance

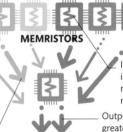

IN 2019, AN **AI PROGRAM** CALLED **PLURIBUS**, BEAT 5 ELITE **HUMAN POKER PLAYERS**

NEURONS

1 Resting state
Nerve impulses pass randomly between a group of neurons – only three are shown here but there could be thousands. Some synaptic junctions send them on easily, others less so. There is no overall pattern and no definite outcome.

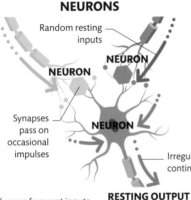

Random resting inputs

NEURON

NEURON

Synapses pass on occasional impulses

NEURON

Irregular activity continues

RESTING OUTPUT

MEMRISTORS

Random resting inputs

INPUT

1 Resting state
A set of electrical memristors receive equal inputs and allow through signals as and when they arrive. Like the neurons, there is no overall pattern and the circuits hardly change.

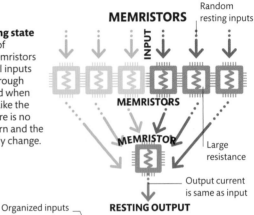

MEMRISTORS

MEMRISTOR

Large resistance

Output current is same as input

RESTING OUTPUT

2 Memory pathway
Recurring, more frequent impulses in specific patterns represent a movement or fact being committed to memory. The repeatedly used synapses boost their connections over time, a characteristic called long-term potentiation (LTP, see pp.26–27 and pp.136–37).

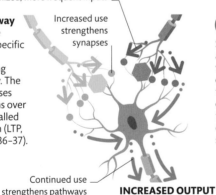

Organized, more frequent inputs

Increased use strengthens synapses

Continued use strengthens pathways

INCREASED OUTPUT

2 Memristor pathway
Stronger inputs arrive at certain memristors, which alter their electrical resistance – the electronic equivalent of LTP. Over time, a recognized pattern develops as the signals strengthen this pathway.

Organized inputs

INPUT

MEMRISTORS

Increased inputs reduce resistance

Output current is greater than input

Continued use strengthens pathways

INCREASED OUTPUT

ELECTRONIC TELEPATHY

Telepathy is the hypothetical direct communication between brains, bypassing senses such as sight. In an experiment using a block-based computer game, instructions to rotate blocks were collected from two players' brains in the form of EEG readings and then communicated, via a transcranial magnetic simulation (TMS) cap, to a third player to make the moves.

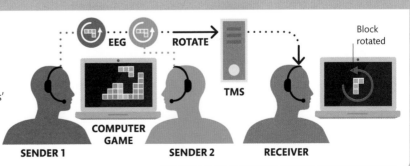

EEG   ROTATE

TMS

Block rotated

SENDER 1   **COMPUTER GAME**   **SENDER 2**   **RECEIVER**

The expanded brain

Medicine uses electrode implants, magnetic fields, radio waves, and chemicals to treat brain problems. These technologies could also potentially enhance normal brain functions.

Boosting the brain

"Overclocking" is the speeding up of a computer's internal clock, which coordinates all its circuits, to push components to work faster and harder. Like computers, the brain uses tiny electrical signals in the form of nerve impulses, which raises the possibility that it might be similarly overclocked. Depending on the region stimulated, this might improve attention and focus, information processing, and memory.

IS IT SAFE TO SPEED UP YOUR BRAIN?

So far, the evidence suggests that tDCS is safe. Thousands of healthy people have taken part in experiments using tDCS and no adverse effects have been noted.

TMS WAND

Wand positioned close to (but not touching) patient's skull

MAGNETIC FIELD

CEREBRAL CORTEX

Negatively charged electrode can inhibit neural activity

Cathode

NANO NEUROBOTS

Researchers are developing almost molecule-sized, robot-like implants, to deliver medical drugs, for example. Next-generation neurobots that are specialized to deliver programmed electronic signals might also accelerate both the way neurons work and how they process their nerve impulses.

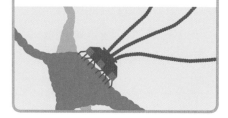

Transcranial direct current stimulation

In transcranial direct current stimulation (tDCS), a direct electric current is passed at a constant low strength through the brain, between padlike electrodes attached to the skin. Sessions of tDCS have helped treat depression and relieve pain. The ability of tDCS to enhance a range of cognitive functions, from creativity to logical reasoning, is being researched. Here, tDCS is shown in use at the same time as TMS, although the techniques are not actually used simultaneously.

Wires form complete circuit

Inhibiting the brain
During cathodal tDCS, the current is negative with respect to the brain's own electrical activity. This has the effect of slowing or inhibiting nerve cells, for example, to reduce hyperactivity.

A **HIPPOCAMPAL PROSTHESIS** CAN IMPROVE **MEMORY** PERFORMANCE BY AS MUCH AS **37 PER CENT**

Transcranial magnetic stimulation

In transcranial magnetic stimulation (TMS), pulses of electric current pass through a coil and generate magnetism that penetrates the skull to influence brain cells and their impulses. The coil's position and motion, and pulse strength and timing, are adjusted to modify particular brain regions. TMS is being trialled for many kinds of brain and behavioural conditions, and also possibly to heighten thinking and other mental processes.

Magnetic pulse
When in use, the magnetic coils change polarity and produce magnetic pulses, which penetrate the scalp. This produces electrical activity in surrounding neurons.

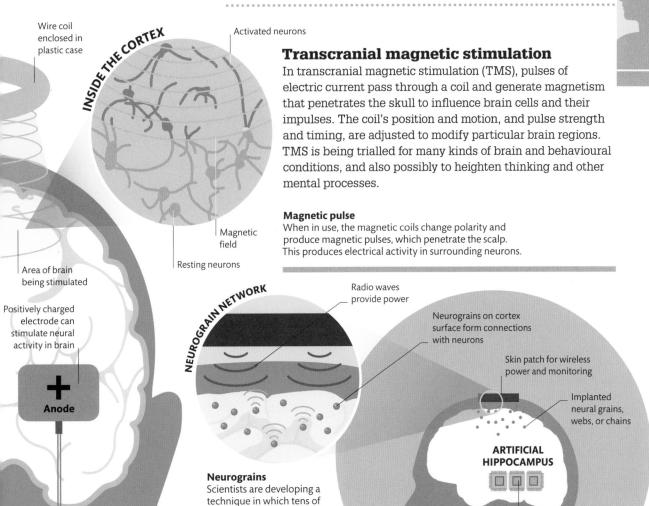

INSIDE THE CORTEX

Wire coil enclosed in plastic case

Activated neurons

Magnetic field

Resting neurons

Area of brain being stimulated

Positively charged electrode can stimulate neural activity in brain

+
Anode

NEUROGRAIN NETWORK

Radio waves provide power

Neurograins on cortex surface form connections with neurons

Skin patch for wireless power and monitoring

Implanted neural grains, webs, or chains

ARTIFICIAL HIPPOCAMPUS

Embedded microprocessor and memory chips

Neurograins
Scientists are developing a technique in which tens of thousands of "neurograins" each independently interface with a single neuron and send data to an electronic patch on the scalp.

Constant electrical current supplied from battery

Stimulating the brain
Anodal tDCS uses a positive current to speed up nerve cell activity. The positions of the skin electrodes determine which brain regions are aroused. Tests show that effects can persist even after the current ceases.

Memory chips

The abilities of electronic devices can be extended by adding more memory, often in the form of microchips. The brain could be similarly upgraded. Microdevices to receive, store, and send data are being shaped like ultrafine webs, chains, and grains. Implanted on or in the cerebral cortex, they could develop connections with individual nerve cells and assist them in thinking and memory. Already, chips can advance hippocampus memory tasks such as long-term recall.

The global brain

Public use of the World Wide Web dates from 1991. Now, the development of a system that may allow our brains to interface with the Cloud is a possibility.

Brain/Cloud interface

Technology is racing to connect human brains into the gigantic electronic network of the Cloud using a brain/Cloud interface (B/CI). A person may eventually be able to access a vast bank of human and electronic knowledge, but many challenges must be overcome. For example, the speed of data transfer must be controlled, or incoming information could be so excessive that it totally overloads our consciousness. And from the start, fully safeguarding each human brain is essential.

Design challenges

Attempting to design a B/CI involves many key elements: a connection to the human brain itself, a method of wirelessly transmitting the brain's neural activity into a local computer network, and establishing how this network interacts with the Cloud.

WHAT IS THE CLOUD?

The Cloud is an immense, worldwide, interwoven network of major electronic equipment. Through it, software and services can run on the Internet, instead of on your computer.

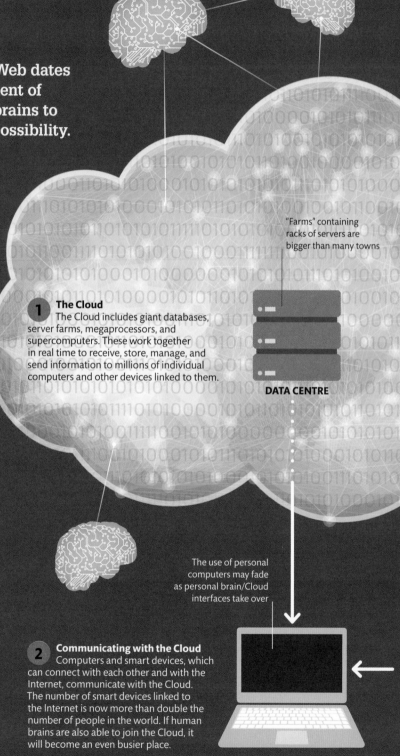

"Farms" containing racks of servers are bigger than many towns

1 The Cloud
The Cloud includes giant databases, server farms, megaprocessors, and supercomputers. These work together in real time to receive, store, manage, and send information to millions of individual computers and other devices linked to them.

DATA CENTRE

The use of personal computers may fade as personal brain/Cloud interfaces take over

2 Communicating with the Cloud
Computers and smart devices, which can connect with each other and with the Internet, communicate with the Cloud. The number of smart devices linked to the Internet is now more than double the number of people in the world. If human brains are also able to join the Cloud, it will become an even busier place.

ACCESS TO THE CLOUD

Deciding which human brains should join the Cloud raises many social and economic issues. Future applications may include faciliating the accuracy of medical diagnoses. But the question of who will be able to use the technology first will have to be considered. Will it be those who need it, those who can best develop it, or those who can pay for it?

NEUROBOTS

Retractable arms work as aerials

Cerebral nanobots
Neurobots implanted in the cerebral cortex, or travelling through its blood vessels with the help of their own micro-positioning guides, act as go-between transmitter-receivers.

Implants may link brain regions as well as to the interface

NEURAL LACE

Scalp skin

Cerebral cortex

Implanted lace unfolds

Cortical intraweb
Neural lace is an ultrafine mesh of electrodes that forms a data-collecting and dispersing area. It also works as a wireless antenna.

3 **Neural implants**
Several technologies are competing to enable early forms of B/CIs. They include neural lace, various types of nanobots, and sub-nanosize particles known as neural dust. Neural dust would allow wireless communication to the brain through microscopic implantable devices inside the body, which are powered by ultrasound.

DISORDERS

Headache and migraine

A dull ache or a sharp or throbbing pain, headache may appear gradually or suddenly and last from less than an hour to several days. Migraine sufferers have episodes of severe headache often accompanied by sensory disturbances, nausea, and vomiting.

Headache is a symptom with a range of possible causes. Probably the most common form of headache is tension headache, in which the pain tends to be constant, in the forehead or more generally over the head. It may be accompanied by a feeling of pressure behind the eyes and/or tightness around the head. It is typically brought on by stress, which causes tension in the muscles of the neck and scalp. This, in turn, is thought to stimulate pain receptors in these areas, which send pain signals to the sensory cortex, resulting in a headache. Another form of headache is cluster headache, which involves relatively short attacks of severe pain.

Migraine headache
Migraine usually occurs over one eye or temple or on one side of the head, although the area of pain can move during an attack. A migraine

typically consists of up to four stages, which vary in intensity and duration (see panel, below). The underlying cause is not known, but research suggests it may be due to a surge of neuronal activity in the brain, eventually stimulating the sensory cortex, resulting in the sensation of pain. Triggers for a migraine include emotional shock or stress; tiredness or lack of sleep; missed meals, dehydration, and certain foods, such as cheese or chocolate; hormonal changes (for many women, migraines are associated with menstruation); and changes in the weather or a stuffy atmosphere.

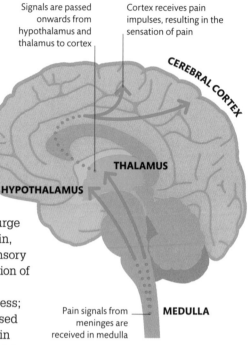

Signals are passed onwards from hypothalamus and thalamus to cortex

Cortex receives pain impulses, resulting in the sensation of pain

CEREBRAL CORTEX

THALAMUS

HYPOTHALAMUS

Pain signals from meninges are received in medulla

MEDULLA

Migraine pathway
When a migraine attack is in progress, pain signals originating in the meninges are transferred to a nucleus in the meninges and then relayed, via the hypothalamus and thalamus, to various regions of the cortex.

IS MIGRAINE A GENETIC DISORDER?

Migraine often runs in the family. Certain genes combine to increase predisposition to migraine, but environmental factors such as stress or hormones are also involved.

MIGRAINE ATTACKS

An attack may begin with an early stage, the prodrome, with symptoms such as anxiety, mood changes, and tiredness or excessive energy. This is sometimes followed by aura, a warning stage that can include flashing lights and other visual distortions; stiffness, tingling, or numbness; difficulty speaking, and poor coordination. The main stage includes severe throbbing headache made worse by movement, nausea and/or vomiting; and dislike of bright light or loud noise. This is often followed by a postdrome stage of tiredness, poor concentration, and persistence of increased sensitivity.

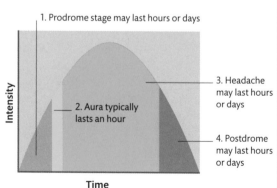

1. Prodrome stage may last hours or days

Intensity

2. Aura typically lasts an hour

3. Headache may last hours or days

4. Postdrome may last hours or days

Time

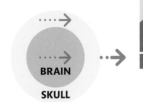

Head injuries

Minor bumps to the head or injuries to the scalp alone have no long-term consequences. However, injury to the brain is potentially extremely serious, and can be fatal.

Direct damage to the brain may occur if both the scalp and the skull are penetrated. Indirect damage occurs as a result of a blow to the head that does not damage the skull. In both cases, head injuries can rupture blood vessels, causing a brain haemorrhage. Minor head injuries typically produce only mild, short-lived symptoms, such as a bruise. In some cases, concussion may follow, and this may cause confusion, dizziness, and blurred vision, which may last for several days. Postconcussive amnesia can also occur. Repeated concussions lead to detectable brain damage, such as impaired cognitive abilities, tremors, and epilepsy.

Severe head injury may produce unconsciousness or coma, and usually brain damage. In nonfatal cases, the effects of brain damage, may include weakness, paralysis, poor memory and/or concentration, intellectual impairment, and even personality changes. Such effects can be long-term or permanent.

1 Rapid movement
When a person is moving rapidly – for example, on a bike or in a car – the skull and brain are moving at the same speed.

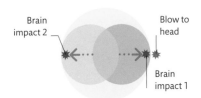

Brain impact 2 · Blow to head · Brain impact 1

2 Stopping suddenly
On impact, the brain smashes into the front of the skull, rebounds, and sustains further injury as it hits the back of the skull.

Epilepsy

Ranging from mild to life-threatening, epilepsy is a brain function disorder in which there are recurrent seizures or periods of altered consciousness, caused by abnormal electrical activity in the brain.

THERE ARE ABOUT 60 TYPES OF EPILEPTIC SEIZURE

Often the cause of epilepsy is unknown, but in some cases it may be due to a brain condition such as tumor or abscess, a head injury, stroke, or a chemical imbalance. Seizures (fits) may be generalized or partial, depending on how much of the brain is affected by abnormal electrical activity. There are several types of seizure. In a tonic-clonic (grand mal) seizure, the body stiffens before uncontrolled movements of the limbs and body begin, lasting up to several minutes. In absence (petit mal) seizures, the victim loses consciousness, although muscle is retained.

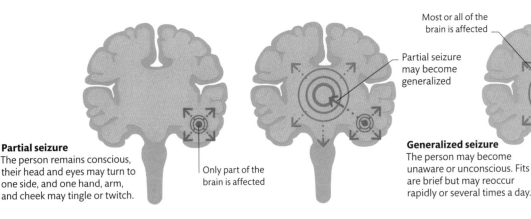

Most or all of the brain is affected

Partial seizure may become generalized

Partial seizure
The person remains conscious, their head and eyes may turn to one side, and one hand, arm, and cheek may tingle or twitch.

Only part of the brain is affected

Generalized seizure
The person may become unaware or unconscious. Fits are brief but may reoccur rapidly or several times a day.

Meningitis and encephalitis

Meningitis and encephalitis are inflammatory diseases caused mainly by infection. Both can produce symptoms such as sudden fever, a stiff neck, light sensitivity, headaches, drowsiness, vomiting, confusion, and seizures.

Meningitis is an infection of the meninges – the membranes that protect the brain and spinal cord, and contain the cerebrospinal fluid that flows throughout the entire nervous system. When infection causes these membranes to swell, the inflammation can ultimately impact every part of the body. Young children whose immune systems are not fully developed are most at risk, although the disease can strike people of any age.

The main cause of meningitis is germs entering the body, whether in the form of bacteria – which can also lead to septicaemia, or blood poisoning – viruses, or fungal infections. However, certain drugs, such as anaesthetics, contain substances that can also irritate the meninges, triggering meningitis.

Encephalitis

Encephalitis is an inflammation of the brain itself, due to an infection or to the immune system attacking the brain in error. A person of any age can contract encephalitis, which can cause serious symptoms such as muscle weakness, sudden dementia, loss of consciousness, seizures, and even death.

Sites of infection
The meninges are the outer dura mater, the middle arachnoid, and inner pia mater. In all forms of meningitis, they become inflamed and impair brain function.

1 MILLION
THE NUMBER OF **PEOPLE WORLDWIDE** AFFECTED BY **MENINGITIS** EACH **YEAR**

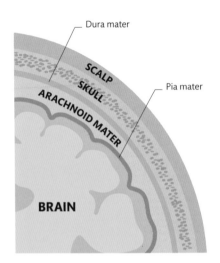

Dura mater

SCALP

SKULL

ARACHNOID MATER

Pia mater

BRAIN

Brain abscess

Also known as cerebral abscesses, brain abscesses are pus-filled swellings in the brain, which often form after an infection or severe head injury that have allowed bacteria or fungi to enter the brain tissue.

The symptoms of a brain abscess may develop slowly or quickly. They can include symptoms such as a localized headache that cannot be relieved by painkillers, neurological problems such as muscle weakness and slurred speech, changes in mental state, high temperature, seizures, nausea, stiff neck, and changes in vision.

Brain abscesses are usually caused by an infection in another part of the skull such as an ear infection or sinusitis; an infection in another part of the body – for example, a pneumonia infection spreading via the blood; or trauma, such as a severe head injury that cracks open the skull.

Assessment and diagnoses of brain abscesses are made via blood tests and a CT or MRI scan. Medication and surgery are the most common forms of treatment.

CONGENITAL HEART DEFECT

A brain abscess can also be a rare complication of a group of conditions known as cyanotic heart disease, which are congenital (present at birth). These cause abnormal blood flow through the heart and lungs, allowing poorly oxygenated blood to be pumped around the body. This oxygen-deprived blood gives affected children's skin a blue, or cyanotic, colour and severely limits their physical activity.

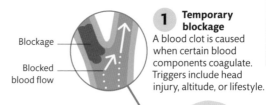

TIA

A transient ischaemic attack, or TIA, is similar to a stroke (see below), which occurs when the blood supply to the brain is interrupted. Unlike a stroke, however, a TIA only lasts briefly.

A TIA is often termed a "mini stroke" and may serve as a warning sign. Indications of a TIA usually disappear within an hour and resemble those found early in a stroke. Symptoms include the sudden onset of weakness, paralysis, or numbness in the face, arm, or leg, typically on one side of the body; slurred speech and difficulty understanding others; blindness or double vision; dizziness or loss of balance or coordination; and a sudden severe headache with no known cause. Depending on the area of the brain involved, symptoms may be similar or different.

Seeking treatment

TIAs most often occur hours or days before a stroke, so it is vital to seek medical attention immediately after a TIA. Roughly one in three people who have a TIA will experience a stroke, and around half of these will take place within a year of the initial TIA.

Blockage

Blocked blood flow

1 Temporary blockage
A blood clot is caused when certain blood components coagulate. Triggers include head injury, altitude, or lifestyle.

Carotid artery supplies blood to brain

Axillary artery

Blood flow resumes

2 Dispersal of blockage
Medication to thin the blood, or surgery to remove the clot, can alleviate a blockage, so blood flows normally.

Blockage disperses

Stroke and haemorrhage

A stroke is a life-threatening condition that occurs when the blood supply to the brain is cut off. There are two main types of stroke: ischaemic and haemorrhagic, and each affects the brain in various different ways.

IN THE **US** SOMEONE HAS A **STROKE EVERY 40 SECONDS**

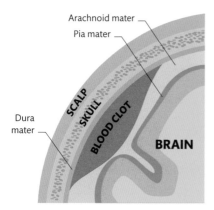

Arachnoid mater
Pia mater
SCALP
SKULL
BLOOD CLOT
Dura mater
BRAIN

Subdural haematoma (haemorrhage)
Bleeding between the brain's protective outer layers, the meninges, forms a clot that puts pressure on the brain, causing a stroke.

If the blood supply to the brain is reduced or interrupted, brain tissue is deprived of oxygen and nutrients. When this happens, brain cells begin to die within minutes. A stroke can be caused by a blockage, usually a blood clot (ischaemic), or when blood spills into the brain or its surrounding tissues (haemorrhagic), often as the result of a ruptured blood vessel or artery.

Symptoms can include slurred speech; paralysis (drooping) or numbness of the face, arm, or leg, which often occurs on just one side of the body; trouble seeing with one or both eyes; and a sudden, severe headache, dizziness, and loss of coordination.

Blood in the brain

Brain haemorrhages can be caused by weak spots in blood vessels that form an aneurysm, or swelling, which bursts, often due to high blood pressure. If this occurs between the two inner membranes surrounding the brain, it is called a subarachnoid haemorrhage. Causes of bleeding within brain tissue (intracerebral haemorrhage) include injury, tumours, or drug use.

Brain tumours

A brain tumour is caused by cells that multiply in an abnormal way. It can occur in any part of the brain, from the intracranial space between the brain and the skull, to deep within the brain itself. Tumours may be benign or malignant, and treatment varies accordingly.

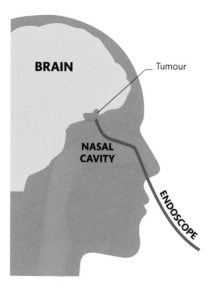

Transnasal brain surgery
Surgeons can now operate on some brain tumours through the nose. The procedure is much less invasive than a craniotomy, where the skull is opened and the brain exposed.

There are approximately 130 different types of brain tumour, and they are classified according to the kind of tumour or the area of the brain in which they grow. Some take years to develop, while others are much faster-growing and more aggressive. Brain tumours can occur in people at any age or stage of life, and the signs and symptoms vary.

Locations and types

The most common types of brain tumour in adults are found in the cerebrum (see pp.28–29). About 24 per cent start in the meninges – the membranes that surround and protect the brain and spinal cord. These tend to be easier to treat, if found early. Around 10 per cent of brain tumours occur in the pituitary gland or pineal glands, which are surrounded by brain tissue.

In children the picture is slightly different. Approximately 60 per cent of childhood tumours occur in the cerebellum or brainstem. Only 40 per cent arise in the cerebrum.

Dementia

Dementia is a term applied to a group of diseases associated with a decline in mental function that occurs most often in adults aged over 65. There are many different types of dementia.

Whether due to reduced blood flowing to the brain, a build-up of protein deposits, or other forms of damage, dementia in all its forms is a progressive disorder. Symptoms typically include mild forgetfulness, which may evolve into apathy or depression, reduced socialization, and loss of emotional control.

In later stages, a person with dementia may lose the ability to be compassionate or feel empathy, or to organize day-to-day activities. People with dementia often become very confused, not recognizing loved ones or knowing where they are. They might hallucinate, have language difficulties, and need help with basic activities such as feeding or dressing themselves.

Diagnosis

While there is no cure for dementia, early diagnosis and treatment can slow the rate of mental decline. Brain scans highlight the areas of the brain most affected in an individual, and treatment can be tailored accordingly. The area most affected in Alzheimer's disease, for example, is the cortex. This part of the brain includes the hippocampus, where new memories are formed.

COMMON CAUSES OF DEMENTIA

Dementia can be caused by various disorders. Some of the most common are listed here.

Alzheimer's disease
A progressive condition in which bodies of protein, called plaques, damage the brain.

Vascular dementia
Impaired blood flow to the brain, such as that caused by stroke, leads to a decline in function.

Lewy body dementia
Protein deposits in the brain's nerve cells affect thinking, memory, and motor control.

Frontotemporal dementia
A form that occurs in the front and sides of the brain, affecting behaviour and language.

Parkinson's disease
Most Parkinson's sufferers develop dementia thought to be related to Lewy bodies.

Creutzfeldt-Jakob Disease (CJD)
Rare, rapid, and fatal, this is caused by an infectious protein called a prion.

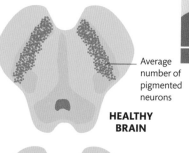

Parkinson's disease

The second-most common degenerative disease after Alzheimer's (see p.50), Parkinson's disease is a neurological disorder that affects movement and mobility by destroying dopamine-producing cells in the substantia nigra, which is located in the uppermost part of the brainstem.

CAN SURGERY BE USED TO TREAT PARKINSON'S DISEASE?

Deep-brain stimulation (DBS) involves surgical implantation of electrodes in the brain that can control, but not cure, the motor symptoms of Parkinson's.

Symptoms manifest gradually, sometimes starting as a mild tremor in one hand. Other signs include muscle stiffness, slurred speech, and a general slowing of mobility. Early stages of the disease usually affect one side of the body, but when 80 per cent of the substantia nigra dies, severe disablement occurs. Late-stage sufferers require assistance with all daily tasks.

Parkinson's mainly strikes adults aged 60 or over and affects more men than women.

Average number of pigmented neurons

HEALTHY BRAIN

Marked decrease in pigmented neurons

DISEASED BRAIN

Changes in the substantia nigra
Parkinson's affects nerve cells in the substantia nigra, which produce the neurotransmitter dopamine. As the cells die, dopamine levels fall, disrupting motor control.

Huntington's disease

Huntington's is a progressive brain disorder caused by a genetic mutation. Early signs include irritability, depression, involuntary movements, poor coordination, and trouble with decision-making or learning new information.

Adult-onset Huntington's is the most common form of the disease, and it usually appears in people in their 30s and 40s. It affects three to seven out of 100,000 people of European origin. Less frequently, it begins in childhood or adolescence, where it causes mobility problems and mental and emotional changes.

Additional symptoms of juvenile Huntington's include slow movements, clumsiness, frequent falling, rigidity, slurred speech, and drooling. Thinking and reasoning abilities are impaired, which affects performance in school. Seizures occur in 30 to 50 per cent of

children with this condition. Juvenile Huntington's disease tends to progress rapidly.

Huntington's chorea

Many people with Huntington's develop involuntary twitching movements known as chorea, which become more pronounced as the disease progresses. They may have difficulty walking, speaking, and swallowing, and may also experience personality changes and a decline in thought processing. Prognosis for people with adult-onset Huntington's is a lifespan of 15 to 20 years after symptoms begin.

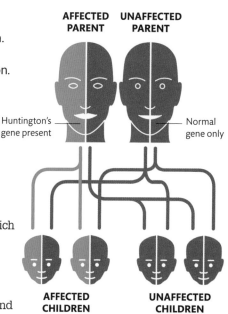

AFFECTED PARENT **UNAFFECTED PARENT**

Huntington's gene present

Normal gene only

AFFECTED CHILDREN **UNAFFECTED CHILDREN**

Patterns of inheritance
Huntington's is classed as an inherited condition. It occurs when a single defective gene is passed on from an affected parent.

Multiple sclerosis

Multiple sclerosis, or MS, is a condition that affects both the brain and the spinal cord. It is believed to be caused when the body's immune system mistakenly damages protective nerve sheaths.

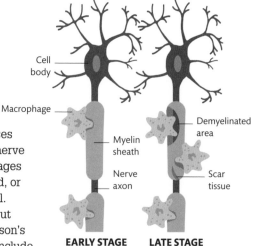

Myelin cells, made of proteins and fats, surround neurons in the central nervous system, enabling messages to travel quickly and smoothly between the brain and the rest of the body. When MS develops, the immune system, which normally fights infection and inflammation, seems to mistake myelin for a foreign body and attacks it with macrophage cells, damaging it and stripping it away. The scars, or plaques, this action leaves behind disrupt impulses normally transmitted along nerve fibres or axons. Neural messages slow down, become distorted, or are simply not delivered at all.

MS may occur at any age but is usually diagnosed in a person's 20s or 30s. Early symptoms include dizziness, vision changes, and muscle weakness. In later stages speech, mobility, and cognition may be affected. The progressive form of the disease results in disability.

Macrophage numbers and MS stages
When MS begins, macrophage cells remove damaged tissue, but also help repair it. In later stages, however, their numbers increase and actually accelerate myelin loss, increasing the severity of symptoms.

Motor neuron disease

Motor neuron disease, or MND, is an umbrella term used to describe a group of conditions that affect motor neurons – the nerves in the brain and spinal cord that tell all the muscles in the body what to do.

PHYSICIST **STEPHEN HAWKING** LIVED FOR **55 YEARS** AFTER BEING DIAGNOSED WITH MND

Genetic, environmental, and lifestyle factors are thought to contribute to the development of MND. Exposure to heavy metals or agricultural chemicals, an electrical or mechanical trauma, military service, or excessive exercise have all been investigated as possible causes, with conflicting results.

Some types of MND, however, do have a genetic basis. Progressive bulbar atrophy, also known as Kennedy disease, results from a mutated gene and affects mainly men. Kennedy disease specifically damages the bulb-shaped lower brainstem, where neurons that control muscles in the face and throat are found.

Whatever their cause, most forms of MND cause symptoms that include general muscle weakness and wasting, cramps, difficulty swallowing, a progressive loss of speech, and limb weakness. Diagnosis includes MRI scans, muscle biopsies, and blood and urine tests.

Although there is currently no cure for MND, symptoms, can be managed to give sufferers the best possible quality of life.

Nerves in dorsal (back) horns carry sensory signals from body to brain

Nerves in lateral (side) horns control internal organs

Nerves in ventral (front) horns control skeletal muscles

KEY
- Ascending tracts carry sensory signals
- Descending tracts control torso and limbs

Spinal cord bundles
Different forms of MND involve different tracts of neurons, located in the dorsal, lateral, and ventral horns of the spinal cord.

Paralysis

The main symptom of paralysis is loss of voluntary control of movement in part of the body. It is classified by the areas of the body affected. Sometimes only one muscle or a small muscle group is affected, but paralysis can also be total, resulting in complete loss of motor function. It can be intermittent or permanently disabling.

Paralysis may affect any part of the body, including the face, the hands, one arm or leg (monoplegia), one side of the body (hemiplegia), both legs (paraplegia), and both arms and legs (tetraplegia or quadriplegia). The body may also become stiff or rigid (spastic paralysis) with occasional muscle spasms, or floppy (flaccid paralysis).

Main causes of paralysis

Paralysis can result from an injury, or be caused by many different disorders, each of which requires specialist assessment. A stroke or transient ischaemic attack (see p.199) can lead to sudden weakness on one side of the face, weakness in one arm, or slurred speech. Bell's Palsy is an abrupt weakness that affects one side of the face, along with earache or face pain.

In addition, severe head or spinal-cord injury can trigger paralysis, while multiple sclerosis or myasthenia gravis – a disease that affects the junction between nerves and skeletal muscles – can cause weakness in the face, arms, or legs that comes and goes. Other causes of paralysis include brain tumours, Guillain-Barré syndrome, cerebral palsy, and spina bifida. Tick-borne Lyme disease causes paralysis that may begin weeks, months, or years after the initial tick bite.

> **WHAT IS THE MOST COMMON CAUSE OF PARALYSIS?**
>
> In the US, the most common trigger is stroke, followed by spinal-cord injuries and multiple sclerosis.

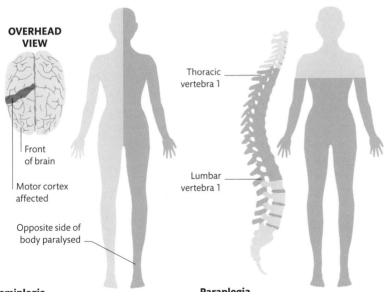

OVERHEAD VIEW

Front of brain

Motor cortex affected

Opposite side of body paralysed

Thoracic vertebra 1

Lumbar vertebra 1

Hemiplegia
Paralysis affects one side of the body, often seen as a result of stroke or brain tumour affecting the motor cortex. Hemiplegia may also be caused by a brain trauma.

Paraplegia
Paralysis affects the legs, and sometimes part of the trunk, usually due to a spinal injury, but it can arise from traumatic brain damage or a medical condition such as a spinal or brain tumour or spina bifida.

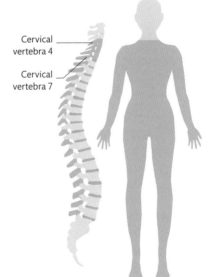

Cervical vertebra 4

Cervical vertebra 7

Quadriplegia
Also known as tetraplegia, both arms and legs are partially or completely paralysed, as is the body from the neck down, usually as a result of a break to the lower part of the neck.

Down's syndrome

Down's syndrome, which affects both physical and mental development, results when an extra copy of a chromosome is made randomly due to abnormal cell division. Babies born with this disorder have identifiable facial characteristics and developmental delays evident from early infancy.

Down's syndrome is also known as trisomy 21, because it creates a third copy of chromosome 21. Experiments conducted on mice have shown that the presence of this extra chromosome disrupts the function of brain circuits involved in memory and learning, mainly in the area of the hippocampus.

The chances of Down's syndrome occurring in a child increase with the mother's age at pregnancy.

Everyone born with it has some level of learning disability. Certain health conditions, such as heart conditions and hearing and vision problems, are more common in people with Down's syndrome.

SCREENING TESTS

Prenatal screening tests such as blood tests and ultrasounds help to predict whether a child might be at risk of Down's syndrome. If the risk is high, these can be followed by two diagnostic tests: chorionic villus sampling and amniocentesis, which analyse fetal cells and amniotic fluid to detect chromosome abnormalities.

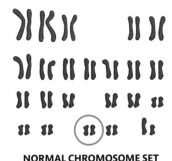
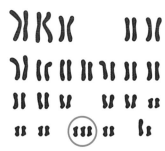

NORMAL CHROMOSOME SET   **TRISOMY 21 CHROMOSOME SET**

Normal and trisomy 21 chromosome sets
Two karyotypes, or photographs of a full set of chromosomes, show a normal male with two copies of chormosome 21 and a male with Down's syndrome, who has three.

Cerebral palsy

Cerebral palsy, or CP, refers to a group of disorders that impair movement, coordination, and cognition. CP is the most common childhood motor disability and is broadly defined as being either congenital or acquired.

Most children are diagnosed with congenital CP, which occurs either before or during birth as a result of brain injury, such as a difficult delivery that deprives the brain of oxygen. Brain infections or a serious head injury, however, can also cause acquired CP more than 28 days after birth.

The nature of CP symptoms depends on the location of the brain damage, but the damage is typically located in the motor cortex, which controls movement.

Symptoms and severity vary enormously and become more evident as a baby develops. Many signs of CP are often not even noticeable in newborns.

Some children with CP have impaired mobility, speech and intellectual abilities, may require a wheelchair, or need support with daily activities. Others may be floppy or rigid, have weak limbs, or trouble walking. Depending on CP type and treatment, affected children live for 30 to 70 years.

TYPES OF CEREBRAL PALSY

CP is categorized by the movement disorder involved. A few types are listed below.

Spastic (or diplegic) CP
People with this type are very stiff, and cannot relax their limbs and muscles. They may walk on their toes or with legs turned inwards.

Athetoid (or dyskinetic) CP
People with this form cannot control various parts of their body and make involuntary writhing or jerking movements.

Ataxic cerebral palsy
Balance and coordination are affected, and there is often loss of voluntary muscle control when using fine motor skills such as writing.

Mixed cerebral palsy
Mixed cerebral palsy involves a combination of symptoms of CP types, due to several damaged motor-control centres in the brain.

Hydrocephalus

Hydrocephalus is a build-up of fluid on the brain, which can damage brain tissue. It is caused by excess cerebrospinal fluid or by fluid not draining away normally. Acquired and normal-pressure hydrocephalus are the two adult-onset forms, but it can also occur in children.

Acquired hydrocephalus is caused by damage to the brain after stroke, haemorrhage, a brain tumour, or meningitis. Enlarged brain cavities fill with excess cerebrospinal fluid (CSF) or block areas where fluid is reabsorbed into the bloodstream.

Causes of other forms

The cause of normal-pressure hydrocephalus is often unknown, but it might be due to underlying health conditions such as heart disease or high cholesterol.

The main symptoms are usually headache, nausea, blurred vision, and confusion.

In children, hydrocephalus can develop after a premature birth, bleeding on the brain, or in cases of spina bifida. In babies and young children, symptoms include a swollen head, but in older children the disorder might show up as severe headaches. Damage caused by the pressure can lead to loss of developmental skills, such as walking and talking.

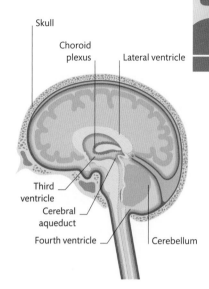

Skull
Choroid plexus
Lateral ventricle
Third ventricle
Cerebral aqueduct
Fourth ventricle
Cerebellum

Fluid on the brain
CSF is created by the choroid plexus, a cellular membrane lining brain ventricles, or cavities. If it isn't reabsorbed, it pressurizes the brain, causing hydrocephalus symptoms.

Narcolepsy

Narcolepsy is a rare, long-term neurological disorder characterized by sudden bouts of sleep. Sufferers are unable to regulate normal sleeping and waking patterns.

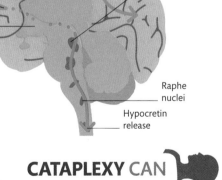

Hypocretin release
Hypothalamus
Locus coeruleus
Raphe nuclei
Hypocretin release

Narcolepsy usually starts around puberty and affects both sexes equally. Symptoms include excessive daytime sleepiness, falling asleep suddenly, and sometimes performing tasks but having no memory of doing so.

The condition can include sleep paralysis – a temporary inability to move or speak, accompanied by terrifying nightmares. Sleep deprivation is a common side-effect.

Cataplexy

Around 60 per cent of sufferers are classed as Type 1, which means they also have cataplexy.

The hypocretin system
Nacrolepsy may be caused by unusually low levels of a brain chemical hypocretin, which is excreted by cells in the hypothalamus. Once released, hypocretin signals neurons in the brain that control wakefulness.

A cataplexic person experiences weakness in muscle control in response to strong emotions such as humour, anger, or pain. There is no loss of consciousness, but sufferers may collapse as a result of loss of muscle tone and are usually unable to speak or move.

CATAPLEXY CAN BE TRIGGERED BY AN EMOTIONAL REACTION SUCH AS **LAUGHTER**

DISORDERS OF UNCONSCIOUSNESS

There are several types of coma, some of which are described here. Some other disorders also show similarities to coma.

Anoxic brain injury
In anoxia, the brain is starved of oxygen. This leads to confusion, agitation or drowsiness, cyanosis (blue-tinged skin, due to low blood oxygen), and loss of consciousness or coma.

Medically induced coma
A drug-induced coma causes a state of deep unconsciousness, which allows the brain to recover from swelling due to stroke or injury.

Locked-in syndrome
Someone with locked-in syndrome is conscious, but brain damage has caused almost complete paralysis. The person almost always communicates using eye movements.

Vegetative state
A person in a vegetative state does not show any meaningful emotional responses, follow objects with their eyes, or respond to voices. Recovery is usually highly unlikely.

Coma

A coma is a prolonged state of deep unconsciousness, whether due to injury or induced to treat a medical condition. Coma patients are unresponsive and may look like they are asleep. Unlike in deep sleep, however, a person in a coma cannot be awakened by any stimulation, including pain.

Comas are caused mainly by head injuries that damage the brain. They often result in swelling, which in turn leads to increased pressure on the brain and damages the reticular activating system – that part of the brain responsible for arousal and awareness.

Bleeding in the brain, a loss of oxygen, infections, an overdose, chemical imbalances, or a build-up of toxins can also trigger a coma, as can the side-effects of various conditions. A temporary and reversible coma occurs in diabetes, for example, when blood-sugar levels remain either extremely high or far too low. More than 50 per cent of comas are related to head traumas or disturbances in the brain's circulatory system.

Treatment

The treatment for coma depends on the specific cause but in general involves supportive measures. Coma patients are placed in an intensive-care unit and may often require full life support until their situation improves.

Depression

More than simply feeling unhappy, depression consists of persistent feelings of sadness, hopelessness, and apathy, accompanied by sleep disorders, fatigue, and appetite changes.

Depression acts on different people in different ways and to varying degrees. Symptoms can be mild to severe (the latter is sometimes referred to as "clinical depression"), and range from constantly feeling unhappy, tearfulness, and a loss of interest in normal activities to an inability to perform daily tasks, and thoughts of suicide.

Physical symptoms

Depression and anxiety often go hand in hand. The disorder may also bring about physical symptoms, such as persistent fatigue, insomnia or excessive sleeping, weight loss or gain, loss of sex drive, and physical pain.

Although it has multiple causes, depression is a genuine illness that can impact all aspects of a person's life. One in ten people have depression at some point in their lives, and it can affect children and adolescents. Depending on its severity, treatment may include medication and psychotherapy.

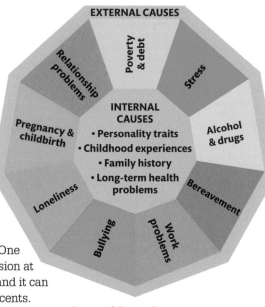

Causes of depression
Stressful life events can act as external triggers for depression. These interact with internal causes that include a family history.

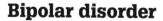

Bipolar disorder

Formerly known as manic depression, bipolar disorder is a mental condition marked by alternating periods of exaggerated elation and depression, in which a person's mood swings suddenly from one extreme to another.

Bipolar phases
People often experience a manic or hypomanic period of feeling high, then a balanced stage of calm followed by episodes of feeling mildly or extremely depressed.

Bipolar mood swings vary enormously, and individuals with the disorder may also have "normal" moods. The patterns are not always the same, however; some people may experience rapid cycling from high to low, or a kind of mixed state.

Treating bipolar disorder involves reducing the severity and number of opposing episodes to give sufferers as normal a life as possible. Medicines such as mood stabilizers, a recognition of triggers and warning signs, psychological treatment such as cognitive behavioural therapy, and lifestyle advice are all used to treat bipolar patients. When effective, episodes usually improve within months.

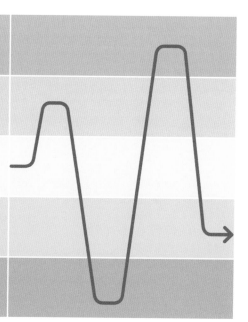

Mania Mania symptoms include euphoria, rapid speech, short attention span, loss of sleep or appetite, and occasionally psychosis.

Hypomania This is a milder version of mania that lasts a few days, often with agitation, and reckless social or financial behaviour.

Balanced mood Euthymia is the term used to describe the relatively stable mood state where a person is neither manic nor depressed.

Mild depression Symptoms may include feeling sad, hopeless, or irritable; a lack of energy; difficulty concentrating; and feelings of guilt.

Depression Emotionally painful, this phase may be marked by flat mood, misuse of drugs and alcohol, self-harm, and suicidal thoughts.

Seasonal affective disorder

Seasonal affective disorder, or SAD, is a depression that comes and goes in a seasonal pattern. It is sometimes known as "winter depression", as that is when symptoms are usually more severe.

The exact cause of SAD is not fully understood, but for those who suffer from winter SAD – where the onset of cold weather triggers symptoms – it is often linked to reduced exposure to sunlight, which limits the functioning of the hypothalamus. This is the part of the brain that controls mood. Some people, however, experience symptoms when warmer weather begins – known as summer SAD.

Other possible causes include a malfunctioning "body clock", which regulates sleep patterns, or abnormally high levels of melatonin.

Symptoms include depression, a loss of pleasure in everyday activities, irritability, feelings of despair, guilt, or worthlessness, and a lack of energy. Tracking symptoms in a diary, exercise, light therapy, and support groups are some self-help methods used by sufferers.

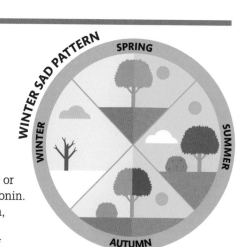

Winter pattern Symptoms begin at the change of autumn to winter, marked by low energy levels and poor mood.

Summer pattern Symptoms reduce or disappear in early spring. There is a return of energy and normal sleep patterns.

Anxiety disorders

Anxiety disorders are a group of mental illnesses characterized by strong feelings of threat and fear, including panic attacks and an inaccurate appraisal of danger. Although there are many types of anxiety disorder, they usually share similar symptoms.

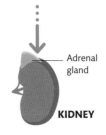

HYPOTHALAMUS

ANTERIOR PITUITARY GLAND

1 In response to stress, the hypothalamus stimulates the pituitary gland to produce adrenocorticotropic hormone (ACTH).

Common anxiety disorders include generalized anxiety disorder, social anxiety disorder, panic disorder, and post-traumatic stress disorder. As well as feelings of fear, physical symptoms are brought on by excessive levels of stress hormones such as cortisol and adrenaline. These symptoms include trembling; sleep problems; cold, sweaty, numb, or tingling hands or feet; shortness of breath; heart palpitations; nausea; and dizziness.

Those with GAD are prone to feelings of intense worry, while panic disorder arises from an extreme bodily response to stress.

People with social anxiety disorder are worried, have an excessively negative self-image, and feel continually observed and judged. PTSD sufferers have feelings of being threatened and constantly on edge, triggered by experiencing or witnessing a traumatic event.

Contributing factors

Many factors influence anxiety disorders, including environmental stress and genetic predisposition; if disorders run in families, they may also be learned. Some may be linked to changes in brain areas that control fear and other emotions.

Adrenal gland

KIDNEY

2 ACTH stimulates production of adrenaline and cortisol by the adrenal glands.

ADRENALINE AND CORTISOL

3 Adrenaline and cortisol trigger various physiological responses, such as a more rapid heart rate and increased muscle tension.

COMMON PHOBIAS	
PHOBIA	**DESCRIPTION**
Arachnophobia	Fear of spiders
Aviophobia	Fear of flying
Claustrophobia	Fear of enclosed spaces
Coulrophobia	Fear of clowns
Mysophobia	Fear of contamination by germs
Necrophobia	Fear of death or dead things
Nosophobia	Fear of developing a specific disease
Trypanophobia	Fear of injections or medical needles

Phobias

An overwhelming, debilitating fear of an object, place, situation, feeling, or animal is known as a phobia. Phobias provoke extreme reactions and involve an unrealistic, intense sense of danger.

A phobia is a type of anxiety disorder characterized by an excessive reaction to a specific trigger. In some cases, just thinking about the threat can make a person feel anxious, a condition known as anticipatory anxiety. Symptoms include dizziness, nausea or vomiting, sweating, palpitations, breathlessness, and trembling.

Phobias can generally be divided into two main types: specific or simple phobias; and complex phobias. Specific phobias centre around a particular object, animal, situation, or activity. Examples include acrophobia (fear of heights) and haemophobia (fear of blood). Common animal triggers for phobias are snakes (ophidiophobia) and spiders (arachnophobia). Simple phobias often begin during childhood or adolescence but tend to decrease in severity over time.

Complex phobias, however, are more disabling. These include social phobia or social anxiety disorder – a fear of social situations.

Obsessive compulsive disorder

Obsessive compulsive disorder (OCD) is a common mental-health condition that affects men, women, and children. A person with OCD experiences repeated intrusive thoughts coupled with a need to perform specific actions over and over in order to relieve any associated anxiety.

OCD can strike at any age but typically develops during early adulthood. It can often be traced to a traumatic event or situation that occurred in childhood or adolescence, and may stem from an out-of-proportion sense of fear, guilt, and responsibility linked to a particular incident.

The obsessive part of OCD is an unwanted and unpleasant fear, thought, image, or urge called an intrusion, which triggers feelings of anxiety, disgust, or unease. The compulsion aspect involves a repetitive behaviour or mental routine that temporarily relieves the intolerable anxiety brought on by the obsession. Both medication and cognitive behavioural therapy (CBT) can be used to manage symptoms.

Genetic factors

About a quarter of OCD sufferers have a family member with the disorder, and studies involving twins suggest that a genetic link is likely. It is also believed that OCD disrupts communication in brain areas, including the orbito-frontal cortex, linked to feelings of reward, and the anterior cingulate cortex, linked to error detection.

TAKES UP AT LEAST
1 HOUR PER DAY

Losing time to OCD
An overwhelming desire to carry out rituals is triggered by anxiety caused by an intrusive thought. This urgent need to count or check objects, wash hands, or repeat thought sequences can use up many hours every day.

Tourette's syndrome

Tourette's syndrome is a complex neurological condition that causes a person to make involuntary sounds and movements called tics. It almost always develops during childhood, usually after the age of two.

Tourette's generally appears in early childhood but before age 15, and is much more likely to affect males than females. Physical tics range from simple blinking, eye rolling, scowling, and shrugging to jumping, spinning, or bending.

The most publicized vocal tic is inappropriate swearing, although in reality this is rare and only affects around one in 10 of those with the disorder. The most usual verbal tics involve making grunting, coughing, or animal sounds.

Tics can cause pain due to muscle strain, and they often increase when a person is stressed, anxious, or tired. Symptoms can change and may improve over time, sometimes resolving completely.

Tourette's tics are often preceded by powerful sensations, like an itch or urge to sneeze. With practice, some sufferers learn to use these cues to control symptoms while in social situations such as school. People with Tourette's may also have OCD or learning difficulties.

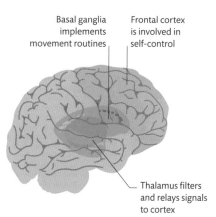

Basal ganglia implements movement routines

Frontal cortex is involved in self-control

Thalamus filters and relays signals to cortex

Implicated brain areas
Tourette's tics are thought to result from an overproduction of the neurotransmitter dopamine, as well as dysfunction in brain areas linked to movement, such as the frontal cortex, basal ganglia, and thalamus.

Somatic symptom disorder

Somatic symptom disorder (SSD) is characterized by an extreme focus on physical symptoms that may or may not be related to an actual diagnosed medical condition. People with SSD, however, truly believe they are ill, and their distress is experienced as bodily, or "somatic" symptoms.

SSD is closely linked to anxiety and depression. Physical manifestations often include pain, weakness, and fatigue; shortness of breath is another common complaint.

Those affected worry excessively about their health and focus on one or several symptoms, even when a medical cause cannot be found for the physical problems they describe. If a diagnosis is found, SSD sufferers are so focused on their conditions that they are often unable to function normally.

Treatment includes antidepressants as well as therapies such as cognitive behavioural therapy (CBT).

Munchausen's syndrome

Munchausen's is caused by severe emotional distress. It is classed as a type of factitious disorder, which is a mental-health condition in which a person acts mentally or physically ill, purposefully fabricating symptoms.

Munchausen's is a rare psychological illness and tends to occur in people who have had traumatic early life events, such as emotional abuse or illness, who have a personality disorder, or who harbour resentment towards authority figures. It is believed to be an extreme form of attention-seeking behaviour. Those affected may tell stories of dramatic occurrences, lie about symptoms, make symptoms worse by deliberately aggravating wounds or ingesting toxins, and even alter test results and falsify records.

A new form of the disorder has been termed Munchausen's by internet, in which a person pretends to have a specific illness and joins an online support group for real sufferers of the disease.

COMMON SYMPTOMS OF FACTITIOUS DISORDERS

Here are some of the symptoms commonly seen in patients with Munchausen's syndrome and other factitious disorders.

A long medical history, often including frequent hospitalization at different locations and visits to several doctors.

Extensive textbook knowledge of the disease reported, as well as of medical practice in general.

A willingness to submit to medical tests, investigations, and even surgery.

An unwillingness to allow medical staff to contact friends and family, or having few visitors when hospitalized.

Many surgical scars or evidence of numerous procedures.

Conditions that get worse for no apparent reason, or which don't respond as expected to standard therapies.

MUNCHAUSEN'S SYNDROME BY PROXY

Munchausen's by proxy is a type of factitious disorder in which carers fabricate or physically induce symptoms of illness or injury in those under their control. Also considered a type of physical and mental abuse, it is usually inflicted on young children by a parent, but sometimes on other vulnerable people under the control of a caregiver, such as an elderly parent being looked after by a son or daughter.

Schizophrenia

Schizophrenia is a mental-health disorder whose symptoms may include delusions and visual or auditory hallucinations. It is a type of psychosis, meaning those affected may not be able to distinguish fantasy from reality.

Schizophrenia can be a difficult disorder to assess. Diagnosis involves examining emotional and cognitive behaviour, and is confirmed by the presence of two or more symptoms that last longer than 30 days. These include disorganized speech or behaviour, catatonia, delusions or hallucinations, and "negative symptoms" such as a lack of emotion or speech.

There are many types of schizophrenia, each with varying symptoms. Paranoid schizophrenics are overly suspicious of others' motives and believe they are being conspired against. A catatonic schizophrenic may withdraw emotionally to the point of seeming to be paralysed, while disorganized schizophrenia includes flat or inappropriate responses and an inability to complete everyday tasks.

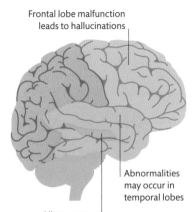

Frontal lobe malfunction leads to hallucinations

Abnormalities may occur in temporal lobes

Hippocampus is usually disrupted

Structural abnormalities
The brains of people with schizophrenia show structural difference in specific areas, such as the frontal and temporal lobes. They also contain less grey matter than normal and this impacts on emotional regulation, motor control, and sensory perception.

DO PEOPLE WITH SCHIZOPHRENIA HAVE A SPLIT PERSONALITY?

The word schizophrenia means "split mind". People with this disorder do not have multiple personalities but instead are cut off from what is real.

1.1%
THE APPROXIMATE PERCENTAGE OF **ADULTS** WITH **SCHIZOPHRENIA WORLDWIDE**

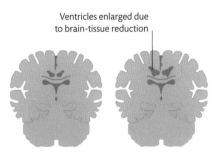

Ventricles enlarged due to brain-tissue reduction

HEALTHY BRAIN   **BRAIN WITH SCHIZOPHRENIA**

Tissue loss
Some schizophrenia patients have enlarged ventricles (the fluid-filled cavities within the brain) as a result of a reduction in brain tissue in surrounding areas.

CAUSES OF SCHIZOPHRENIA

Despite years of research, the causes of schizophrenia remain unclear. It may be linked to genetics, brain chemistry, life experiences, drug use, prenatal or birth trauma, or a combination of these.

Genetics
About 80 per cent of people with schizophrenia show a hereditary predisposition to the disorder. However, genes are not the sole cause, as environmental factors and family history are also considered relevant.

Brain abnormality
MRI studies of the brain show reduced grey matter in several regions, including the prefrontal cortex. This area is important for emotion regulation, decision-making, and complex cognitive tasks such as efficient planning.

Brain chemistry
Two brain chemicals, glutamate and dopamine, are linked to schizophrenia. Elevated dopamine levels may cause hallucinations. Low glutamate levels may trigger psychotic episodes, while high levels damage brain cells.

Environment
A predisposition to developing schizophrenia can be triggered by fetal exposure to a virus, birth trauma, or malnutrition. Environmental triggers include extreme stress, family relationships, or use of mind-altering drugs.

Addiction

Addiction stems from a chronic dysfunction of a brain system that regulates reward, motivation, and memory. A person suffering from an addiction craves a substance or behaviour, often with no concern at the time about the consequences of pursuing it.

An addiction involves the repeated use of, or engagement with, a substance or activity for feelings of pleasure. Psychological and social symptoms include many behaviours, such as lack of self-control, obsession, and risk-taking. Common physical symptoms are changes in appetite, appearance changes, sleeplessness, injury or disease caused by substance abuse, and increased tolerance to the source of the addiction, so that more and more if it is required to achieve the same amount of pleasurable reward. Removal of the addiction source causes reactions such as sweating, trembling, vomiting, and behavioural changes.

Chemical pleasure

Addiction affects the brain's structure as well as how it functions. Humans feel excitement and pleasure when the brain releases neurotransmitters like dopamine, followed by a feeling of intense satisfaction from hormones such as endorphins. Endorphins relieve stress and pain in ways similar to drugs such as cocaine.

For many people, creative or physical activities, such as playing a musical instrument or exercising, release enough neurotransmitters to provide pleasure and satisfaction. For others, however, certain drugs, alcohol, and risk-taking activities such as gambling induce a quicker and much more extreme form of pleasure before eventually disrupting and damaging normal neurotransmitter circuitry.

Such artificial stimuli flood the brain with dopamine, then create feelings of intense satisfaction once endorphins are released. The resulting "high" is registered by the hippocampus as a long-term memory, which leads to an urge to repeat the experience. Once this desire overrides normal behaviour and the ability to function, it is classed as an addiction.

TO WHAT EXTENT IS ADDICTION INHERITED?

Studies involving twins and adopted individuals show that about 40–60 per cent of susceptibility to an addiction is inherited.

Why people are susceptible to addiction is not fully understood, but evidence suggests that genetic make-up may be a factor in some cases. Genes, after all, dictate not just how we respond to substances, but what reactions occur when those substances are withdrawn. This may explain why some people become more readily dependent on alcohol, for example, than others.

Evaluating individuals for a suspected addiction includes the use of diagnostic tests as well as psychological assessments. They are then referred to specialists for treatment and rehabilitation.

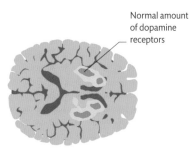

Normal amount of dopamine receptors

HEALTHY BRAIN

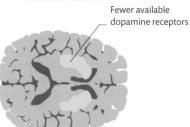

Fewer available dopamine receptors

COCAINE USER

Cocaine use and dopamine
Using cocaine reduces the availability of receptors for the neurotransmitter dopamine. The result is that, over time, the user has to consume more of the drug to achieve the same sensation of reward.

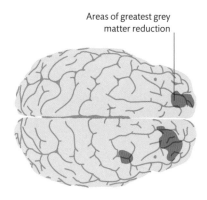

Areas of greatest grey matter reduction

Grey matter and methamphetamine
The use of methamphetamine shrinks the amount of grey matter in the brain's frontal cortex, among other areas, leading to a decline in mental function.

Personality disorder

Individuals who display persistent inappropriate, inflexible, or extraordinary behaviours, or have problems relating to others have a personality disorder, or PD. There are several PD types, ranging from antisocial (BPD) to schizotypal, but some sufferers can manage their lives without medical help.

A personality disorder involves a consistent pattern of behaviour that deviates noticeably from that which is considered acceptable by society. Symptoms usually appear by adolescence and can lead to long-term difficulties for sufferers, in terms of navigating relationships and simply functioning effectively in social situations.

The many types of PDs are broadly grouped into three groups or "clusters": suspicious; emotional and impulsive; and anxious (see panel, below). Each type has its own symptoms. For example, a person with a suspicious personality disorder is typically antisocial, easily frustrated, and has difficulty controlling anger. Borderline personality disorder – a type of emotional and impulsive PD – is associated with disturbed ways of thinking, impulsive behaviour, and problems controlling emotions.

The anxious cluster includes avoidance personality disorder, which is characterized by feelings of inadequacy and extreme sensitivity to negative criticism and rejection. Unsurprisingly, people who have this type of PD also experience severe social anxiety.

The PD brain

Some people affected by PDs have an unusual amygdala, part of the limbic system – the most primitive part of the brain that regulates fear and aggression. People who have PDs involving excessive levels of fear generally have smaller amygdalae than those who do not, and the smaller the amygdala, the more overactive it seems to be. In addition, the hippocampus,

75 PER CENT OF PEOPLE DIAGNOSED WITH BPD ARE WOMEN

which also helps to control emotions, is often reduced in the brain of individuals with PDs.

People with PDs usually find that talking therapies help them to gain a better understanding of their thoughts, feelings, and behaviours. Therapeutic communities, a form of group-therapy treatment, can also be effective but require a high level of commitment. Medication may also be used in some cases to control depression and anxiety.

PERSONALITY DISORDER CLUSTERS		
CLUSTER A: SUSPICIOUS People with these PDs tend to be considered odd or "eccentric". They fear social situations and have problems relating to other people, whom they view with a great deal of suspicion. Some sufferers appear detached, others introverted.	**CLUSTER B: EMOTIONAL AND IMPULSIVE** These PD types are characterized by a lack of emotional control. Cluster B individuals often bully or manipulate others, are self-centred, and prone to dramatic, excessive displays, forming intense but short-lived relationships.	**CLUSTER C: ANXIOUS** The most fearful cluster of PDs. Those in this group are generally anxious, submissive to others, and have difficulty coping with life on their own. They tend to be oversensitive, inhibited, extremely shy, or perfectionists.
Paranoid	Antisocial	Avoidant
Schizoid	Borderline	Dependent
Schizotypal	Histrionic	Obsessive-compulsive
	Narcissistic	

Eating disorders

Eating disorders are emotional mental-health problems that include an extreme relationship with food. Most revolve around an obsessive focus on weight and body shape, which can damage health and may even be life-threatening.

Although they can occur at any life stage, eating disorders usually develop among adolescent and young-adult age groups. The three most common types are anorexia nervosa (or simply anorexia), bulimia nervosa (bulimia), and binge-eating disorder (BED) (see panel, below). Diagnosis involves psychological evaluation as well as physical examinations, such as blood tests and measuring the person's body mass index (BMI).

Anorexia always involves weight loss, and a very low BMI is usually flagged in diagnosis. Those affected by both bulimia and BED do not tend to have a low BMI, and may be slightly overweight. Eating-disorder symptoms include a preoccupation with weight and body shape, avoiding food-based activities, eating very little or overeating then purging (self-induced vomiting), extreme use of laxatives, and exercising too much. Sufferers may also have stomach problems, an abnormal weight for their age and height, menstrual problems or disruption, dental issues, sensitivity to cold, fatigue, or dizziness.

Underlying factors

The causes of eating disorders are not fully understood, but those affected are more likely to have a family member with a history of eating disorders, depression, substance misuse, or addiction. Social pressure and criticism may contribute to a focus on eating habits, body shape, or weight. Some occupations, such as ballet-dancing, acting, sport, or modelling, where there is a focus on being slim, are likely to have a higher number of people with eating disorders than other professions. People with eating disorders may also suffer from anxiety, low self-esteem, perfectionism, and sexual abuse. Treatment includes nutritional education, psychological or talking therapies, and group programmes.

1. The person eats large amounts of food rapidly, often in secret, and may go into a kind of dazed state while doing so.

2. Anxiety drops as eating temporarily numbs stressful, sad, or angry feelings.

3. Low mood returns, bringing with it self-loathing and disgust, due to guilt and shame associated with bingeing.

4. Anxiety rises as eating provides only short-term relief from psychological pain. Depression sets in.

5. Thoughts of food become more and more dominant, as distressing feelings increase.

6. Need to binge-eat becomes urgent; the person often buys special food for the purpose.

The bingeing cycle
Those with a binge-eating disorder use food to numb emotional pain instead of addressing its psychological cause positively. The result is a destructive cycle.

TYPES OF EATING DISORDER	
DISORDER	**DESCRIPTION**
Anorexia nervosa	Mainly affects young women. Involves an obsessive desire to maintain a low body weight by eating little and over-exercising.
Bulimia nervosa	Bingeing and purging occur in this disorder. The body weight is often normal, but bulimics possess a severely negative self-image.
Binge-eating disorder	Regular excessive eating, usually planned and consumed rapidly and in secret, is followed by intense guilt and shame.

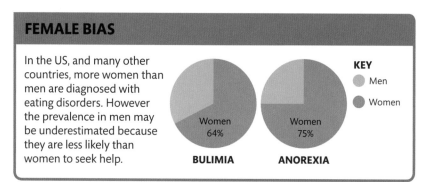

FEMALE BIAS

In the US, and many other countries, more women than men are diagnosed with eating disorders. However the prevalence in men may be underestimated because they are less likely than women to seek help.

KEY
- Men
- Women

Women 64%
BULIMIA

Women 75%
ANOREXIA

Learning disabilities and difficulties

A learning disability is a sign of impaired cognitive abilities and is reflected in a person's general intelligence or IQ. Learning difficulties do not affect IQ levels but make information-processing harder. Both affect how a person acquires knowledge, masters new skills, and communicates.

An intellectual or learning disability occurs when brain development is affected in some way, whether through injury or a genetic abnormality. Learning disabilities range from mild and moderate to severe and profound. The most severe may even mean that an affected person will face problems coping with independent life skills.

Specific causes include genetic mutations such as in Down's syndrome, or fetal head injuries, maternal illness, a lack of adequate oxygen to the brain before or during birth, or brain damage from a childhood illness or injury. Some

conditions have no identifiable cause. No two learning disabilities are the same, and they can include a wide variety of symptoms.

Some people with learning disabilities can talk easily and care for themselves but may take longer than usual to learn new things. Others may not be able to communicate at all. Some may also face mobility problems, heart defects, or epilepsy, which can shorten life expectancy.

Affected people may also have associated learning difficulties – for example, someone with cerebral palsy (see p.204) may have impaired cognitive function and dyspraxia, or a person on the autistic spectrum may have a severe form of developmental delay.

HOW COMMON ARE LEARNING DISABILITIES?

An estimated one to three per cent of the world's population has some form of learning disability, and people in low-income countries are the most affected.

Learning difficulties

Distinguishing some learning disabilities from learning difficulties can be challenging. Generally, however, learning difficulties do not affect intellectual ability or aptitude but instead impact on how the brain processes data. Someone with dyslexia, for example, which makes reading, writing, and spelling difficult, often has dyspraxia, which affects fine motor skills and coordination.

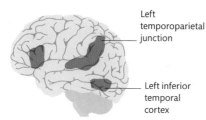

Left temporoparietal junction

Left inferior temporal cortex

NORMAL READERS

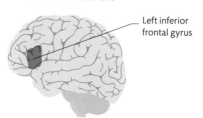

Left inferior frontal gyrus

DYSLEXIC READERS

The dyslexic brain
Areas of the brain activated during reading differ hugely in normal readers and dyslexics. Only the left inferior frontal gyrus activates in dyslexics, but this is paired with increased activity in the right hemisphere – which is why many dyslexics are highly creative.

SOME COMMON LEARNING DISABILITIES AND DIFFICULTIES	
NAME	**DESCRIPTION**
Dyslexia	Impaired ability to learn to read and/or write. As well as poor reading and spelling skills, dyslexics may also have problems with sequences, such as date order, or difficulties organizing their thoughts.
Dyscalculia	Difficulty processing numbers, learning arithmetical concepts such as counting, and performing mathematical calculations. Dyscalculia often occurs alongside dyslexia or other learning difficulties.
Amusia	Literally meaning "lack of music", amusia is sometimes known as tone deafness and means that a person with normal hearing is unable to recognize musical tones or rhythms, or reproduce them.
Dyspraxia (developmental coordination disorder)	The inability to make skilled movements with accuracy, dyspraxia is often first noticed in childhood as "clumsiness". It can cause problems with establishing spatial relationships, such as positioning objects.
Specific language impairment	Indicated by a delay in acquiring language skills where no developmental delay or hearing loss is present, specific language impairment has a strong genetic link and often runs in families.

Attention deficit hyperactivity disorder

Inattentiveness, hyperactivity, and impulsiveness are the main symptoms of the mental-health disorder known as attention deficit hyperactivity disorder, or ADHD. It usually appears in early childhood, but symptoms may increase from the ages of six to 12 and persist into adulthood.

The main symptoms of ADHD include impetuosity, difficulty concentrating, a "short fuse", disorganization, prioritization issues, trouble multitasking, and being extremely active or restless. While attention deficit disorder (ADD) shares similar symptoms, ADD sufferers are less hyperactive, and their main problem is an inability to concentrate.

ADHD symptoms can improve with age, but many adults who have been diagnosed with the condition as a child may continue to experience problems throughout their lives. Such difficulties often become evident in the workplace, where an employee has to comply with routines and rules; in this scenario, a person with ADHD may perform less well than would normally be expected.

Additionally, people with ADHD may also experience additional problems, such as sleep and anxiety disorders.

What causes ADHD?

Because ADHD is a developmental problem that appears to run in families, researchers suspect that there is some genetic basis for the disorder. If genetic faults are to blame, they are likely to be complex and involve more than one gene. The condition has been linked to fetal impairment caused when a mother smoked or drank alcohol while pregnant. Being born prematurely or coming into contact with toxins such as lead in early childhood can also trigger ADHD.

People with ADHD often have learning difficulties (see p.215), although these are not necessarily linked to intelligence or ability levels. Research has revealed biological and structural differences, including a smaller size and lower blood flow, in the brains of people with ADHD compared to those of people without it. Some studies show that brain chemicals such as dopamine may be lower than normal in those with ADHD.

CAN CHANGES IN DIET HELP PEOPLE WITH ADHD?

Some parents report behaviour spikes after certain foods are eaten, but there is no clear evidence that ADHD is caused by diet or nutritional issues.

MEN ARE **THREE TIMES MORE LIKELY** THAN WOMEN TO BE DIAGNOSED WITH **ADHD**

SYMPTOMS OF ATTENTION DEFICIT HYPERACTIVITY DISORDER		
HYPERACTIVITY Hyperactivity is the term used for someone who is abnormally or extremely active. A hyperactive person is very restless, easily distracted at school or work, and often cannot still for more than a few seconds or minutes at a time.	**INATTENTIVENESS** Inattentiveness is associated with ADHD. It is defined by behaviours such as a lack of focus, failure to notice the needs of others, or being preoccupied and not capable of giving sustained attention to the matter at hand.	**IMPULSIVITY** Impulsivity is characterized by actions carried out without any forward planning or awareness of immediate or future consequences. Impulses can be related to emotional situations and physical activity, and can seem to be involuntary.
Difficulties sitting still	**Concentration difficulties**	**Frequently interrupting**
Constant fidgeting	**Clumsiness**	**Inability to take turns**
Talks more loudly than others	**Easily distracted**	**Talking excessively**
Little or no sense of danger	**Poor organizational skills**	**Acting without thinking**
	Forgetfulness	

Autism spectrum disorders

Autism spectrum disorders (ASD) is a term used to describe a group of developmental problems, all of which are characterized by communication and behavioural difficulties. The word "spectrum" refers to the wide variety of types and severity levels of symptoms experienced by people with ASD.

People who have ASD find it hard to interact and communicate with others. They also tend to have restricted interests and repetitive behaviours, and are often more or less sensitive than others to light, sound, or temperature. This causes them to retreat into themselves.

ASD occurs in people at all levels of intellectual ability and is most often diagnosed in the first two years of life. It is a lifelong condition. Physical symptoms may include repetitive body movements, such as pacing, rocking, or hand-flapping.

Communication problems

Children with ASD may have language difficulties, and some start talking relatively late in life. Their tone of voice might be very flat, very fast, or singsong. About 40 per cent of children with ASD don't talk at all, and between 25 and 30 per cent develop some language skills during infancy but then lose them later in life.

SYMPTOMS OF AUTISM SPECTRUM DISORDERS	
SYMPTOM	**DESCRIPTION**
Social communication	ASD affects social communication because the development of language is impaired. Verbal and non-verbal social communication problems include difficulties interpreting social situations, identifying social cues, and blunt or inappropriate conversational interactions.
Repetitive behaviour	People with ASD often engage in repeated activities, such as hand-flapping, body-rocking, or may harm themselves by continuous biting or skin picking. They may also exhibit body-twirling, or other complex body movements, along with rituals such as counting or arranging objects.
Focused interests	Those with autism often think in very black-and-white terms, with an intense focus on specific interests or obsessions. These can range from spinning objects to collecting birthdates or identifying flight paths.
Sensory sensitivity	Some type of sensory processing problem is usually (although not always) related to a diagnosis of ASD. Those affected may be over- or under-sensitive, and experience difficulties with smell, taste, sight, hearing, touch, balance, eye movement, and body awareness.

High-functioning adults with ASD may be successful in academic fields, yet have difficulty with practical and social skills, such as understanding social cues. Most seem blunt, cannot lie, and may focus obsessively on one aspect of life, such as cleanliness.

Social awkwardness is usually accompanied by social anxiety. Other symptoms of ASD include a highly acute awareness of noise, smell, touch, or light, and extreme food preferences.

ASD sufferers who have intellectual disabilities may show a high aptitude in other areas such as having a photographic memory or numerical ability; however, sometimes the disability is so profound that those affected cannot speak meaningfully, engage in self-harm, and need daily care.

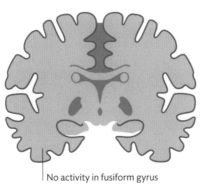

| Activity in fusiform gyrus | | No activity in fusiform gyrus |
| **NORMAL BRAIN** | | **AUTISTIC BRAIN** |

ASD and normal brain comparisons
Those with ASD find it hard to process faces. In a non-autistic person, activity shows in the temporal lobe's fusiform gyrus, where recognition occurs. In the austistic brain there is no such corresponding activity.

Index

Page numbers in **bold** refer
to main entries

Acknowledgments

DK would like to thank the following people for help in preparing this book: Janet Mohun and Claire Gell for helping to plan the contents; Helen Peters for compiling the index; Joy Evatt for proof-reading; and Katy Smith for design assistance.

Senior DTP Designer Harish Aggarwal

Jackets Editorial Coordinator Priyanka Sharma

Managing Jackets Editor Saloni Singh

The publisher would like to thank the following for their kind permission to reproduce or adapt graphs and brain images:

(Key: a-above; b-below/bottom; c-centre; f-far; l-left; r-right; t-top)

46 Data from the American Academy of Sleep Medicine: (bl). **50 PNAS:** Based on Fig. 1 from "A snapshot of the age distribution of psychological well-being in the United States", Arthur A. Stone at al., Proceedings of the National Academy of Sciences Jun 2010, 107 (22) 9985-9990; DOI: 10.1073/pnas.1003744107 (bl). **51 APA:** (Excluding explanatory annotation): Based on Fig. 2 - Longitudinal estimates of age changes in factor scores on six primary mental abilities at the; latent construct level. From "The Course of Adult Intellectual Development" by K. W. Schaie 1994, American Psychologist, 49, pp. 304-313 © 1994 by the American Psychological Association (br). **59 PNAS:** Based on Fig. 2A from "Sex differences in structural connectome", Madhura Ingalhalikar et al., Proceedings of the National Academy of Sciences Jan 2014, 111 (2) 823-828; DOI: 10.1073/pnas.1316909110 (crb). **103 PLoS Biology:** Based on Fig. 4 from "Grasping the Intentions of Others with One's Own Mirror Neuron System", Iacoboni M, Molnar-Szakacs I, Gallese V, Buccino G, Mazziotta JC, Rizzolatti G, Feb 2005 PLoS Biol 3(3):e79. doi:10.1371 / journal.pbio.0030079 (crb). **155 PLoS ONE:** Based on Fig. 3A from "Neural Substrates of Interactive Musical Improvisation: An fMRI Study of 'Trading Fours' in Jazz", Gabriel F. Donnay, Summer K. Rankin, Monica Lopez-Gonzalez, Patpong Jiradejvong, Charles J. Limb, Feb 2014 PLoS ONE 9(2): e88665. https://doi.org/10.1371/journal.pone.0088665 (bc).

For further information see:
www.dkimages.com